W9-CEO-942

The PROBLEM *of the* SOUL

Other Books by Owen Flanagan

Dreaming Souls: Sleep, Dreams, and the Evolution of the Conscious Mind (2000)

Self-Expressions: Mind, Morals, and The Meaning of Life (1996)

Conciousness Reconsidered (1992)

Varieties of Moral Personality: Ethics and Psychological Realism (1991)

The Science of the Mind (1991)

The PROBLEM *of the* SOUL

TWO VISIONS OF MIND AND HOW TO RECONCILE THEM

OWEN FLANAGAN

A Member of the Perseus Books Group

Published by Basic Books,
A Member of the Perseus Books Group

Designed by Jeffrey P. Williams
Set in 11.5-point Weiss by the Perseus Books Group

A CIP catalog record for this book is available from the Library of Congress
ISBN 0-465-02460-2

02 03 04 05 / 10 9 8 7 6 5 4 3 2 1

To Alasdair MacIntyre
and to the memory of
Francisco Varela
dear friends and teachers

CONTENTS

PREFACE

This book is about the conflict between two grand images of who we are: the humanistic and the scientific. The humanistic image says that we are spiritual beings endowed with free will—a capacity that no ordinary animal possesses and that permits us to circumvent ordinary laws of cause and effect. The twentieth-century philosopher Roderick Chisholm sums up the main idea this way: When we act freely we exercise "a prerogative which some would attribute only to God: each of us, when we act, is a prime mover unmoved. In doing what we do, we cause certain things to happen, and nothing—or no one—causes us to cause those events to happen." The scientific image says that we are animals that evolved according to the principles of natural selection. Although we are extraordinary animals we possess no capacity that permits us to circumvent the laws of cause and effect. The question is this: Which is it? The two images, at least as depicted in these terms, are incompatible. The answer can't be both. Or, if it is, there is a lot of explaining to do.

We want to see ourselves truthfully, and we also want our stories to depict life as if it really means something. But we live in a world in which two distinct self-images, vying for our allegiance, disagree about human nature and about the ground of meaning. One image says humans are possessed of a spiritual part—an incorporeal mind or soul—and that one's life and eternal fate turn on the state of this soul. The other image says that there is no such thing as the soul and thus that nothing—nothing at all—depends on its state. We are finite social animals. When we die, we—or better: the particles that once composed us—return to nature's bosom, not to God's right hand.

The humanistic image, embraced by most laypersons, scientists, and intellectuals, claims to be uplifting and inspiring. We create ourselves by exercising our free will. If we will well, when we die we reap eternal reward. From the perspective of the scientific image, this idea is extremely implausible, excessively flattering, and self-serving. If it provides meaning, it does so at cost to the truth.

But the scientific image, from the humanistic perspective, is dehumanizing—it drains life of meaning. Life has no transcendent purpose, and the quest to live morally becomes just one among several quirky features of our kind of animal. Defenders of the scientific image claim that their image need not be seen as depressing or inhospitable to a dignified, moral, and meaningful life. Humanists are skeptical. It is part of the humanistic perspective to deem science, especially the mental and human sciences, as a threat. Science is reductive and materialistic, and it offers no resources to help us find our way in the high-stakes drama that is life.

Perhaps the truth about human nature is eternally incompatible with an uplifting story about the meaning of life. The truth can sometimes be painful. Perhaps honestly acknowledging the truth about human nature would necessarily undermine any sense of purpose and meaning, and bring ennui and nihilism in its train. We tell our children stories about Santa Claus and tooth fairies. These stories are false, but they please the kids. Could we be acting like grown-up children who fabricate false stories for our own comfort, for the sake of meaning? Possibly. Some defenders of the scientific image think this is exactly the case.

But there is another possibility. Perhaps the mythic stories we are used to telling about our nature are beloved not because they are indispensable to a meaningful life but only because we have been historically conditioned to think so. Perhaps, too, there is sufficient room in the scientific image for mind, morals, and meaning that it can preserve much of what it means to be a person. If this is so, then the stark inconsistency between the two images is—at some level at least—more apparent than real. This is what I think.

In my experience, most defenders of the scientific image either ignore the dominant humanistic image or deem it silly and misguided, while defenders of the humanistic image simply assert that the scientific image is de-meaning. But both images share a common aspiration: to maintain a robust conception of what it means to be a person, a being possessed of consciousness, with capacities for self-knowledge and the ability to live rationally, morally, and meaningfully. No advocate of the scientific image has yet made an adequate effort to explain carefully, patiently, and explicitly how the scientific image can do this. That is the task I set for myself.

"The problem of the soul" is a shorthand way of referring to a cluster of philosophical concepts that are central components of the dominant humanistic image. These concepts include, for starters, a nonphysical mind, free will, and a permanent, abiding, and immutable self or soul. It is the survival of these concepts that ordinary people fear are at risk from scientific progress, and this fear is at the root of the deep-seated resistance to the scientific image. Ordinary, intelligent people have a (somewhat inchoate) view that nothing less than the meaning of life turns on how these concepts fare. If the nonphysical mind, free will, and the soul are not real things but are mere appearances, then, well, it is the end of the world—at least the end of the world as we know it.

Some readers will undoubtedly claim that "the soul" is an old-fashioned term, now rarely used save in specifically religious contexts, and then only by true believers. But in my experience, the soul is a concept that most people, even so-called secular humanists, believe in or implicitly assume. A quick and reliable diagnostic test to show this is to ask a person if they believe humans possess free will. Nearly everyone will answer "yes." Then ask what their free will is or consists in. This test reliably reveals that most people believe they have the ability to circumvent natural law in something like Chisholm's sense: "each of us, when we act, is a prime mover unmoved. In doing what we do, we cause certain things to happen, and nothing—or no one—causes us to cause those events to happen." The trouble is that the only device ever philosophically invented that can do this sort of job is an incorporeal soul or mind. This is why the soul is the problem.

When people are introduced to the conflict between the humanistic and scientific images, Copernicus and Galileo, Darwin and Freud are mentioned as the major players. And they are. Furthermore, each hit to the humanistic image—the cosmological, the biological, and the psychological—advanced by these great thinkers has in various ways been accommodated without causing that image to completely unravel. The standard tactic is to acknowledge that we are both fallen and part animal, and to admit that the theories of Copernicus and Galileo and Darwin and Freud enrich our understanding of our animal part while deflating certain excessively flattering but ultimately unnecessary stories about our nature and place in the universe. Thus both the humanistic and scientific images are given room to coexist within a certain circumscribed space. The humanistic image reveals our spiritual nature; science unlocks the secrets of the external world and our animal essence.

This sort of accommodation cannot work, however, without the premise that we are only *partly* animal. But now the scientific image—especially as it is being propelled by advances in evolutionary biology, genetics, and the sciences of the mind, cognitive science and cognitive neuroscience, in particular—asserts that we are *all* animal. Furthermore, it rejects what Gilbert Ryle called "the myth of the ghost in the machine," or what Antonio Damasio calls "Descartes' Error," the humanistic belief that the mind or the soul interacts with—but is metaphysically independent of—the body. The mind or the soul is the brain. Or better: Consciousness, cognition, and volition are perfectly natural capacities of fully embodied creatures engaged in complex commerce with the natural and social environments. Humans possess no special capacities, no extra ingredients, that could conceivably do the work of the mind, the soul, or free will as traditionally conceived. Clearly, the tense rapprochement between the two images cannot abide this sort of challenge.

Can we do without the cluster of concepts that are central to the humanistic image in its present form—the soul and its suite—and still retain some or most of what these concepts were designed to do? My answer is "yes." Although there are no souls or nonphysical minds, no immutable selves, although there is no such thing as free will, there are still, at the end of the day, persons. What is a person? A conscious social animal that deliberates, reasons, and chooses, that is possessed of an evolving or continuous—but not a permanent or immutable—identity, and that seeks to live morally and meaningfully.

So what's lost? What's different? What is lost is the cluster of philosophical concepts that many believe are central to the humanistic image. My own view is that since these concepts don't refer to anything real, we are best off without them. Furthermore, once they are eliminated, we still have more or less what most people desire—a picture of persons as rational social animals. Mind, morals, and the meaning of life are all still in place.

An attentive reader will rightly worry: What about God and personal immortality? The philosophical concepts that I say must go not only support a certain traditional conception of the person. They also support and are supported by a mother lode of views in philosophical theology, views about God, his nature, and our place in his plans.

There is no point beating around the bush. Supernatural concepts have no philosophical warrant. Furthermore, it is not that such concepts are displaced only if we accept, from the start, a naturalistic or scientific vision of things. There simply are no good arguments—theological, philosophical, humanistic, or scientific—for beliefs in divine beings, miracles, or heavenly afterlives. Of course, I can't just say this and expect the reader to let me get away with it. I'll do my best in this book to show why supernatural concepts lack rational warrant.

But my main aim is to show that the scientific image can give us pretty much everything we can sensibly want from the concept of a person. Most of what we traditionally believe about the nature of persons remains in place even without the unnecessary philosophical concepts of the soul and its accompanying suite. Furthermore, the moral and communal functions of religion need not lose their meaning. Even if the concepts of the soul, God, and the spiritual afterlife go the way of phlogiston, the attempt to awaken the mind and heart by words and hymns that gesture at that which is higher and greater than us is not rendered senseless. There is much that is greater than each of us taken alone. There is love and friendship. There is benevolence and compassion expressed by a feeling of connection to all creatures, indeed even to the awesome inanimate cosmos. And there are certifiably great individuals who exemplify our most noble aspirations—Moses, Confucius, Buddha, Jesus, and Muhammad among them. Words such as "holy," "God," and "grace" gesture metaphorically but not literally. As modes of metaphorical expression, they may continue to advance what is noble and good.

If this is not enough, if you feel that you really *need* the concepts of the soul and its mates to refer to real things, then you want more than any philosophically respectable theory can provide.

We are immensely complex animals, possessed of remarkable gifts—consciousness and abstract thought, as well as the ability to live rationally, morally, and meaningfully. But we are animals through and through. Unless we accept this truth about ourselves we will lack self-knowledge of the most basic sort. And we will be tempted to puff up and decorate our basic human image with all manner of magical and supernatural props, all manner of self-serving nonsense about the ways in which God—or various other major and minor spirits—bless (sometimes exclusively) our kind of person.

In order to show that the scientific image leaves ample room for a humane and dignified concept of the person, I will need to wade waist-deep into the waters of philosophical theology. One reason is that the problem of the soul—that is the entire class of troublesome concepts that pertain to the nature of persons and their minds—has deep theological roots. Indeed, one might say that the concept of the soul is a theological concept, and thus that I can't extract the concept of souls while leaving persons fully intact without explaining the various ways the traditional concepts, of persons and their minds are part of a theological world-picture. The reconciliation I aim for can only succeed if I can provide a convincing case for the claim that the concepts of person, mind, and volition can survive cutting them loose from this theological world-picture.

The word *person* comes from the Latin *persona,* which originally referred to the mask worn in Greek and Roman drama. It eventually took on a broader meaning—part, character, role represented by an actor, eventually citizen. It is tempting to think of persons as characters in a play and of human life as a drama. Drama has structure, plot, theme, and meaning, and it brings both thought and the emotions powerfully into play.

We want human life to possess all of these characteristics—structure, plot, theme, meaning. We want it to enliven our hearts and minds. In the drama of real life we not only want to be ourselves, we want both our character and the larger plot to be genuinely meaningful, not simply the result of a good plot and good performance.

We draw meaning from the large-scale stories, pictures, and images we have of our nature and our place in the world. Most people find these images in churches, synagogues, and mosques. The stories told in these places leak out and permeate our ideas of what it means to be a person, what it means to live a moral life, and where and how fulfillment is to be found.

I can put the point another way: Most people are religious, and even many who aren't, are in a way they may not realize. When people in the West talk about mind, morals, and the meaning of life, we either explicitly acknowledge that we embed them in a religious philosophy, or use these ideas and concepts in ways that reveal that they are so embedded.

This is why we cannot adequately understand the humanistic image, even in its alleged secular variant, without acknowledging the deep theological roots that help give humanistic concepts their sense. Many people explicitly believe that we are made in God's image after all.

I hope to show that we can preserve much of what we mean when we speak of "mind," "soul," "the self," and "free will" without continuing to endow them with that part of their meaning that comes from their religious or theological roots. These concepts can be tamed and naturalized without giving us a depressing picture of what it means to be human. But this can only be done by applying a more thoroughgoing honesty about the way such concepts still draw on their religious roots.

The scientific image and its proponents no longer accept the restricted role of understanding and explaining our animal "side." Our animal side is our only side. We are all animal and the brain is our soul. This is not such bad news. There are still persons. Consciousness exists. Love, friendship, and morality all remain. Nothing disappears, save for certain fictions that never existed to begin with. My aim is to provide a naturalistic philosophy of human nature compatible with the scientific image that retains what is beautiful, true, and inspiring in the humanistic image. The project to live morally and meaningfully expresses our most noble aspiration. But this project proceeds inauthentically if it is built, as it now is, on a false picture of the kind of creatures we are.

The PROBLEM *of the* SOUL

ONE

Human Being

> Do I dare disturb the universe?
>
> —*T. S. Eliot, "The Love Song of J. Alfred Prufrock," 1917*

PERENNIAL PHILOSOPHY, the wisdom of the ages, provides an unfolding humanistic story of our nature. In the West, we look to the writings of Plato, Aristotle, Augustine, Aquinas, Descartes, Hume, Kant, and Hegel, as well as to the Old and New Testaments, to discover both who we are and how we ought to live. These works give us an image of what it is to be a person and guide our way along life's path—they tell us the way to goodness, righteousness, knowledge, happiness, and, most importantly, a meaningful life.

Literature and art, especially the literature and art we call "great" and "timeless"—the works of Sophocles and Shakespeare, Michelangelo and Leonardo, Beethoven and Bach—embody, reinforce, and revitalize perennial philosophy. True wisdom, unlike most everything else, withstands the ravages of the ages. It withstands not only time and indifference but radical challenges from the likes of Protagoras, Thrasymachus, Callicles, Julien Offroy de La Mettrie, Friedrich Nietzsche, Karl Marx, and all manner of radical or minor poets and artists. The perennial philosophy and classic literature, art, and music constitute the canon, the body of texts and artworks that reveals the widely accepted truth about humans, our location in the greater scheme of things, and our duties to our fellows and to lesser and greater beings.

Science has lately gotten into the business of trying to understand human nature in completely naturalistic terms. Before the nineteenth century, scientists were busy explaining inanimate nature as well as animate bodies. Astronomy, physics, and plant and animal physiology proceeded apace. Inquiry into the workings of the material world carried the blessing of perennial philosophy, as did its discoveries, albeit with certain notable exceptions. But persons, or at least their souls and minds, were off limits to science, and for a principled reason. Minds and souls, not being physical, were not a proper object of scientific study.

Things have changed. In the last one hundred and fifty years, the possibility that persons might be wholly understood scientifically—and this by virtue of the fact that we are completely natural beings, fully embodied, and possessed of no immaterial parts—has become both feasible and likely. Among those who have in different ways encouraged the scientific understanding of persons are Charles Darwin, Emile Durkheim, Karl Marx, Max Weber, Sigmund Freud, Ivan Pavlov, B. F. Skinner, Margaret Mead, E. O. Wilson, Richard Dawkins, Daniel Dennett, Jane Goodall, Frans deWaal, Alan Turing, John von Neumann, Noam Chomsky, Francis Crick, James Watson, Charles Sherrington, Patricia and Paul Churchland, and Antonio Damasio.

Some of these thinkers display a reductive, scientistic streak—B. F. Skinner, who boldly proposed that we get "beyond freedom and dignity," springs to mind—others do not.[1] But even the nonscientistic folks on the list, for example, Goodall, deWaal, and Damasio, believe that humans are animals who do not house souls. Oliver Sacks could also make the list. Indeed putting him on it might be a good idea since Sacks writes about his patients in a way so filled with love and recognition of their humanity that I suspect it goes unnoticed—as it does to a lesser degree with Goodall, deWaal, and Damasio—that Sacks is a thoroughgoing physicalist. He describes his patients as fully human and with a deep and abiding desire to retain their humanity. But they don't, as Sacks conceives them, have souls. Sacks may be the perfect model—the possibility proof, as it were—of a neuroscientist who has

[1]I use the term "scientism" to depict a family of attitudes that center on the idea that everything worth saying can be said in scientific language. Hardly anyone really thinks this. But some philosophers and scientists suggest that the humanistic image gets things so wrong that it is beyond repair—that concepts such as mind, freedom, and dignity are outdated and that we are better off without them. One variant of scientism says that what is really real is the world as depicted by particle and quantum physics or string theory; everything else is mere appearance. Rob Wilson calls this view "smallism."

no patience for reductive, scientistic description or explanation, but who nonetheless sees humans as animals.

Desouling Persons

There is no consensus yet about the details of the scientific image of persons. But there is broad agreement about how we must construct this detailed picture. First, we will need to demythologize persons by rooting out certain unfounded ideas from the perennial philosophy. Letting go of the belief in souls is a minimal requirement. In fact, desouling is the primary operation of the scientific image. "First surgery," we might call it. There are no such things as souls, or nonphysical minds. If such things did exist, as perennial philosophy conceives them, science would be unable to explain persons. But there aren't, so it can. Second, we will need to think of persons as part of nature—as natural creatures completely obedient and responsive to natural law. The traditional religious view positions humans on the Great Chain of Being between animals on one side and angels and God on the other. This set of beliefs needs to be replaced. There are no angels, nor gods, and thus there is nothing—at least, no higher beings—for humans to be in-between. Humans don't possess *some* animal parts or instincts. We *are* animals. A complex and unusual animal, but at the end of the day, another animal.

It is no surprise that the images offered by perennial philosophy and science conflict, even though many of the details of the scientific image are not yet in place. Perennial philosophy teaches that we are spiritual beings and that everything turns on perfecting our spiritual nature. The scientific image is committed to desouling us. If there is room for human perfection within the constraints of the scientific image—and I think there is—it cannot be spiritual perfection as traditionally conceived, because we do not possess spiritual components.

As the third millennium opens, some are trying to diminish the conflict between the humanistic image and science by assigning them different domains and roles. No modern version of the Councils of Constantinople, in which various popes and emperors met to define church doctrine and condemn challenging heresies between the fourth and fourteenth centuries, has negotiated this wary standoff.

The terms of this unspoken cease-fire call for a division of labor between the humanistic image as defined by perennial philosophy and science. The human sciences reveal our animal nature—something perennial philosophy has always acknowledged is there—but they have nothing to say about our full nature and place in the cosmos, conceived along the lines of the Great Chain of Being. From the point of view of those inspired by perennial philosophy (and this includes most scientists and philosophers, as well as ordinary intelligent folk), proponents of the scientific image, as I have described it, are playing out of their depth, behaving like the "know-it-all" adolescents many of us remember being—never lacking in confidence, often loud and brash, but seriously deficient in humility and wisdom. Predictably, the defenders of the scientific image in turn regard the defenders of the humanistic image as harmless but somewhat annoying old farts.

Since in fact the human sciences cannot deliver a picture remotely as detailed as that offered by perennial philosophy—the latter having had, among other things, a three millennium head start—defenders of the humanistic image can easily perceive science as akin to an annoying advertisement that makes incredible promises and is worth ignoring.

When I speak of the missing details of the scientific image, I don't just mean the small details, such as which neurotransmitter fixes memories in the brain, or the medium-size details such as whether the mind is best described as running one general-purpose computer program or numerous special-purpose ones, but the big details as well: nature versus nurture, for example, or whether we are sneaky egoists like our chimpanzee relatives or warm, cuddly, and largely peaceful like our bonobo relatives. All three levels of details are missing at present. Without them, the scientific image is more of a scheme for an image than a richly detailed picture or map. When it overreaches, it is seen in the eyes of detractors as just a "philosophy" in the pejorative sense—a mere theory or an advertisement for one.

Some might think that this ideological standoff is, under these circumstances, a sensible one. But I don't think so, for three reasons.

First, the scientific approach is now delivering some of the goods that it has promised. It becomes harder each day to dismiss the scientific discoveries that are in fact yielding new knowledge about human being." Martin Heidegger used the neutral word *dasein*—which means "being" or literally "there-being"—to indicate that our understanding of human being, of the nature of persons, and of what and how we are, is still uncertain. Heidegger was not one smitten by science, but in seeking a neutral word to speak of our way of being, he acknowledged that the scenario offered by perennial philosophy is deficient—at once sketchy and incomplete despite its long life, excessively flattering and thus inauthentic, and possibly logically incoherent in places. He was right. When we examine perennial philosophy with the standards of logic and evidence that it has developed and encourages us to use, we see that it fails to produce a robust and authentic picture of human being.

Second, the conflict between the humanistic image—as refined and endorsed by perennial philosophy—and the scientific image affects the lives of ordinary people, even if they are unaware of it. This conflict is not like many others between competing scientific or philosophical hypotheses, which can be resolved in the halls of academe and whose eventual resolution may not significantly affect how ordinary people think about themselves or the world. We live in the space of images and conflict between these images has consequences for how we live. Many people have felt Dostoyevsky's disquieting worry—"If there is no God, then everything is allowed." If the perennial philosophy were entirely secure, if it were firmly held and believed with assurance, this is not a thought one could even have. To be sure, this thought is more likely to arise from the news that there are Nazis among us than that there are neurons within us—more likely to arise, as it did for Dostoyevsky, as well as Nietzsche, Marx, and Freud, from noticing that disordered persons and politics abound than that science dispenses with God. But the fact is that the cumulative discoveries of the human sciences in the last century and a half, the combined forces of psychology, sociology, anthropology, primatology, evolutionary biology, genetics, and neuroscience, significantly affect the way ordinary people think and feel about themselves. And many have noticed that the human sciences appear to have better resources to explain how there could be Nazis among us than does perennial philosophy.

Third, the scientific image has developed methods to sort out most of the multifarious triggers and mechanisms that make us tick. Mind science, especially neuroscience, is fast maturing. Why does that make such a difference? Because there is no longer any place for the soul to hide. The mind, as conceived by perennial philosophy, has been toppled off the pedestal from which it supposedly performed its magic through mysterious capacities of free will to transform and manipulate—possibly to override—the combined forces of the natural and social environment, genes, and whatever else came its way.

We now live in the age of mind science. The first President Bush dubbed the 1990s "the decade of the brain" in expectation of dramatic research breakthroughs and medical applications. Prior to this growth in brain research the brain was largely *terra incognita*. I graduated from college in 1970 and considered going to graduate school to study the human mind. For me this meant studying the human brain. But in those days that pretty much meant observing the behavior of rats in mazes before and after removing or destroying various brain parts. I hate rats, but that aside, the whys and wherefores of rats confused about mazes hardly seemed like a promising approach to understanding the human mind. In 1970, behaviorism still reigned in many quarters, and behaviorists rightly referred to the mind/brain as a "black box."

Now everything is different. For the first time in history, the proponents of the scientific image can see into the black box. Cognitive science and cognitive neuroscience have reliable methods and tools for examining and identifying the way the mind/brain works. It is not as if we know in remotely complete detail how the mind/brain works. But we know this much: The mind/brain does its magic through the operation of neurons, with axons and dendrites that form synaptic connections, and via electrical and chemical processes that mediate attention, remembering, learning, seeing, smelling, walking, talking, love, affection, benevolence, and gratitude. René Descartes, a thinker as attuned as anyone could be in the seventeenth century to working out a picture of persons that is compatible with science, famously believed that we think with immaterial minds. He was wrong: The brain working in concert with the rest of the nervous system is our *res cogitans*—our thinking stuff. We are fully embodied creatures. Genes, culture, and history work through and with this extraordinarily complex tissue to make us who we are.

It is not as if we now understand human nature. However, what we can say with confidence is that we have a good sense of what needs to be explained—human nature and behavior—and of how to go about explaining it. There are many unknown forces in genes, mind, and culture that will affect the story we eventually tell about what it means to be a person, about why we think, feel, and behave as we do. But no scientifically minded person thinks we will need resources beyond those available to genetics, biology, psychology, neuroscience, anthropology, sociology, history, economics, political science, and naturalistic philosophy to understand the nature of persons. The beliefs in immaterial minds, and minor and major spirits, are in need of explanation. But spiritual forces will not do any explaining. Unless God and the angels are tampering with us, unless astral forces and extraterrestrials are messing with our brains and perceptions, science will one day be enough to provide a true picture of our *dasein*. Or—better, synthetic scientifically inspired philosophy will do so. This new philosophy will modify and, where necessary, displace and replace the perennial philosophy.[2]

---

[2]A synthetic scientifically inspired philosophy must replace perennial philosophy because the latter is inadequate. There are many ways a philosophy can be inadequate. It can be vague, it can contain falsehoods, or the falsehoods it contains can be troublemakers. My view is that the perennial philosophy in the West does not suffer particularly from vagueness. Its details have been pretty well worked out. But it does contain falsehoods, for example that we possess nonphysical minds, free wills, and souls that constitute our essence, and that we have prospects for immortality; these falsehoods are troublemakers. Yet the fact remains that the perennial philosophy that supports the image of a souled and potentially heaven-bound humanity does help many of us to locate meaning for our lives and to find our way morally. This partly accounts for its resiliency. The synthetic scientifically inspired philosophy that I believe will eventually replace perennial philosophy must encourage ethics and the quests for meaning, enlightenment, even perfection, however defined. Such a philosophy might well take inspiration from the parts of perennial philosophy that do these things well. Thus when I speak of "replacement," one might think in radical terms of throwing out the old and bringing in the new, or one might think of something kinder and gentler, like keeping the beloved sputtering old car but with a rebuilt engine. As best I can, in this book I try to perform a kind, gentle replacement—one that keeps much that works well and is beautiful in perennial philosophy. But the replacement must still be perceived as radical, for it claims that certain central claims of perennial philosophy are false and troublemakers to boot.

So the wary standoff between humanistic and scientific perspectives cannot be—in fact, is not being—maintained. Mind science is exploding, and in concert with the other human sciences it is overturning the traditional conception of mind. The mind is nothing like what perennial philosophy says it is. Something—the humanistic image as endorsed by perennial philosophy, it seems—has to give.

Human Animals

Many think that the conflict between the perennial philosophy and the scientific image, and especially the resistance the scientific image provokes, lies in our resistance to materialism, physicalism, naturalism. Call it what you will. The rub comes from denying two things: what there is, and all there is, is the natural world; and that we are animals. According to the scientific image, we are conscious animals living at a certain time in the broad sweep of natural and social history. Most traditions—religious, cultural, political—say we are more than animals. This is not true. What we are is unusual animals. We have fancy capacities to reflect on our motives, intentions, and behavior and to modify, redirect, and control ourselves. We are rational animals and we are animals that make things—*Homo sapiens* and *Homo faber.*

One reason we resist the scientific image is that we have trouble truly seeing ourselves as we are—as a kind of animal who knows, feels, reasons, and creates. For centuries, we have pictured ourselves as something we are not, as God's chosen creature standing above animals but beneath the heavenly host. This is wrong. We are, I repeat, animals. But it is hard to really see ourselves as animals, for several reasons. First, the nature of any thing, a human being included, is not easy to grasp. We are animals who can know things, but our own nature is hardly transparent to us. It took us tens of thousands of years to discover that water is H_2O, and the nature of water, as things go, is simple. Our natures, on the other hand, are very, very complicated. Second, we are story-telling animals. We make sense of things through stories, and stories, especially when bundled together, generate grand pictures. We picture ourselves and our world through stories, grand stories.

Many thinkers—Nietzsche and Freud spring to mind—have supposed that we humans have difficulty seeing ourselves as we truly are because of a powerful tendency to want to picture ourselves in the most flattering way possible, and this requires dishonesty. I have a different diagnosis: It is exceedingly difficult to know the truth about our natures because we are extraordinarily complex. Many have thought that, as self-conscious creatures, they could just do a sufficiently disciplined "look-see" and *voilà*—know what they are. It doesn't work that way. If human nature was transparent to us by virtue of being human, it might have worked. But we are not self-transparent. Meanwhile, we tell what Rudyard Kipling called "just-so" stories about our natures, our minds, our souls, akin to the sort of fable one might make up about how the leopard got its spots. A just-so story, once repeated and believed, becomes an established story that tells us of our origin, our fate, who we are and why. When we stick to this story, we feel secure about our place in the larger scheme of things. When someone tells a different story it is disquieting and threatening and destabilizing. It feels awkward, unbecoming, untrue.

If this is right, then we can think of Nietzsche's and Freud's insights about flattery and disguise in a somewhat different way. It is not so much that we, by virtue of being human, require false and flattering stories, but rather that we settle into the best stories we can find. Stories build incrementally on prior stories, tale upon tale. What makes for a flattering story is determined not by some absolute standard of the most flattering story, the most flattering image, but by standards that have evolved within the stories themselves. The story we come to accept sets the terms of what we believe is true, normal, and good about us. Conformity to the story—both to the image of human nature it presents and to the norms and ideals of a good or attractive person—sets the standards for a flattering self-image.

Although the core self-image of an age is normally long-lived, we may usefully compare it to passing fashion. Certain decades are remembered for their fashion. Consider the look of the roaring twenties or the disco look of the seventies. Each looked flattering in its time, but soon the fashion faded and the look became dated. In retrospect fashions often seem silly.

At the start of this new millennium, as in the last millennium, and even more so in the previous one, conceiving of persons as made in God's image, as composed of body and soul, and as acting outside the constraints of nature's laws is in. It is flattering to us. Occasionally, radical thinkers suggest a different image: Spinoza, La Mettrie, Nietzsche, Freud, Darwin, Skinner, pick your favorite. Their images were unflattering. What they offered at the time was not in fashion.

Once the story is in place that says we sit above animals but below angels and God, the alternate story that says we are animals seems (and is) unflattering. But it is unflattering only in contrast to the story that says we humans are not animals, and that it is unflattering to be compared to animals, let alone to be one. To be sure, there is a certain conception of animals—that they are dirty and dumb, for example—that is unsavory when applied to persons. But there is no reason to apply this negative perception of animals to ourselves (indeed there is no reason to think of uncaged animals this way). Even though the scientific image conceives of us as animals, it does not deny that we are incredibly smart—*Homo sapiens*—nor that we create things—*Homo faber*—including hygienic tools and technologies. We were never a stupid animal and many of us now have ways of keeping ourselves clean.

We could, I claim, get used to thinking of ourselves as animals, as fully embodied, and thus as soulless and without prospects for an afterlife in the company of spirits, without experiencing discomfort or fear at the very idea. But it will not be easy. A full century and a half after Darwin, most people, including most intellectuals who pay homage to the idea that we are animals, don't really accept that we are animals. The problem is that we have not yet found a way to tell a story about what it means to be a human animal that isn't in some way disquieting. This has as much to do with negative views about what it means to be an animal than with what it means to be a person. It certainly is not based on any objective fact that being an animal is inherently unseemly. If we were to pay more than lip service to the idea that we are animals by devoting careful and sustained attention to the kind of animal we are, I believe we could get used to the idea and embrace it. This shift will require adjustments, including disposing of various myths, illusions, and delusions about our natures, but a frank and honest understanding of ourselves as animals needn't produce existential horror or unremitting nausea. At least, not in perpetuity.

Being Ought to Mean Something

Beyond our difficulty in accepting that we are animals—material beings living in a material world—there is another reason for our discomfort with the scientific image. It is hinted at when people announce that Darwinian evolution renders life meaningless. Many say that evolution is ateleological, that it removes purpose from the universe. *Dasein* has no purpose. That which has no purpose has no meaning. This lies at the heart of nihilism.

We need to be careful here. What Darwin showed was how the evolution of species is unplanned. There is no need to posit an intelligent designer who created *Homo sapiens* and the nearly unnamable variety of nature's bounty (Darwin quietly acknowledged that his theory undermined traditional theology). Nowadays it is widely accepted that a Big Bang launched our universe—a single colossal crunch of matter and subsequent explosion. A billion years passed, during which galaxies and planets congealed and fell into great whirling patterns; later, water, ammonia, nitrogen, oxygen, and a few other ingredients combined to yield primitive unicellular life-forms that interacted with the physical environment in ways that eventually produced multicellular life, and the rest, as they say, is history.

Does the theory of evolution by natural selection imply that life has no meaning or purpose? No. It just says that species aren't designed by intelligence or for a purpose. Evolution produced intelligence. We are an example. But evolution does not require intelligence to yield intelligence. The biblical view is that the existence of intelligence requires an Intelligent Designer. But evolution demonstrates how intelligence arose from totally insensate origins.

Evolution also produced meaning. We are language users and words and signs have meaning. But evolution does not require that meaning be born of meaning or that meaning be produced by a Meaner. The Mother of All Meaning—whatever that may mean—did not possess meaning or act from a wellspring or reservoir of meaning. Likewise, evolution produced our many emotions, but no Emoter on

High was directing the action. We are sexy, but no conception I am familiar with thinks God is.

A common mistake is to think that evolution drains the universe of all meaning and purpose. This belief is completely erroneous, sometimes willfully so. According to the theory of evolution by natural selection, intelligence, meaning, and purpose exist, as surely as blue crabs and sandbar sharks. Evolution is a process through which life-forms, behaviors, and consciousness emerge. It is the best theory—one backed up by a preponderance of evidence—we have of how these things came to be. Humans are very smart and resourceful. Our thoughts, words, and actions have meaning to us, and most of us aim to live in purposeful, meaningful ways.

The theory of evolution might truly be said to drain the world of transcendent or ultimate meaning, but it does so only if meaning is equated with or required to rest on intelligent design or the activities of a God. Evolution says there is no need for a creator God who stands outside the universe, organizing it, endowing it with meaning and purpose, and perhaps also planning for our heavenly afterlife—once, that is, we've shed our nonessential animal part.

This idea will have to go. Our animal part is our only part. Once we shed it, once we die, we are gone.

Is this bad news? I don't see why. Is it bad news that, insofar as our lives have meaning and purpose, we have to find and make our meaning and purpose and not have them created and given to us by a supernatural being or force? Again, I don't see that it is. It seems like good news that meaning and purpose are generated and enjoyed by me and the members of my species and tribe, rather than imposed by an inexplicable and undefinable alien being.

Of course, not everyone will see things this way. For all of human history we have been gripped by all manner of illusion and nonsense about meaning and purpose. Even though no one knows what ultimate or transcendent meaning would be if there was such a thing, we like whatever it is these words and images gesture toward. It seems comforting. But, again, there is nothing authoritative about the story set out and promoted by perennial philosophy, which tells us both that there is a ground of ultimate or transcendent meaning and that it is only by virtue of this ultimate or transcendent meaning that anything else can have any meaning at all. Meaning, says this story, only comes from meaning. If human life is to have any meaning at all, it must be assigned by some powerful being—a higher, insubstantial one—who created the meaning and either put it in our heads or scattered it about for us to find. This is a misguided view, but it is a big part of the traditional story in whose terms we have long seen ourselves. And boy do we like our stories.

The Big Problem Is Ethics

But I am underestimating one source of our unease with the scientific image. One specialty of the perennial philosophy is ethics, "the science of ought," which gives guidance to how we ought to be. Here's the rub. Science is in the business of description, explanation, and prediction. It tells us how things are, why they are as they are, and what we can expect to happen. But science does not traffic in oughts. Or so it seems.

If science really does not traffic in oughts, then the scientific image does not remotely drain the world of meaning and purpose but teaches us that meaning and purpose are in our hands. Still, insofar as living a meaningful life involves finding meaning and purpose, it requires that we make certain discoveries and decisions about what is right and wrong, good and bad, and what leads to happiness, flourishing, and contentment—what Aristotle called *eudaimonia*. This requires that we think about what we ought to do to live well. The wisdom of the ages is chock-full of excellent ethical advice. Love thy neighbor; Thou shalt not kill; Maximize happiness; Abide the categorical imperative and never treat another as mere means to an end; The unexamined life is not worth living; Love and friendship are among the greatest goods, so seek them.

Science, at least as normally done, does not provide ethical wisdom. In fact it seems to have no resources whatsoever for generating ethical opinion. Empirical observation tells us that some people are loving and compassionate and others are egoists. Perhaps science can tell us certain things about why they are so. But there is nothing that comes from the application of science, as normally understood, that tells us one ought to love one's neighbor. While a scientist might say you ought to be loving and compassionate, she does not get this idea from science, nor can she justify it in scientific terms.

Science tells us how to manipulate genes, how to end life painlessly, how to abort a fetus, and how to make and deliver nuclear weapons. But science does not give any guidance about whether we ought to use this knowledge, and if so under what circumstances.

So here's the dilemma. One might think, as I do, that a philosophy inspired by the scientific image will beat the part of perennial philosophy that endorses the humanistic image hands down when it comes to explaining what human being is, but that it will fare miserably in helping us decide how we ought to be and what we ought to do. Any good philosophy will need to offer wisdom about who we are as well as about how we ought to be. Since a philosophy inspired by the scientific image has no resources to do the latter, it can provide no ethical wisdom at all. Thus it is not helpful as a philosophy on how to live well.

This is an important point. Indeed, I think the inability of science to provide ethical wisdom is as much responsible for our resistance to the scientific image as what science says about us being animals or about there being no transcendent ground for meaning and purpose. Scientists speak out on important ethical issues, but they do not ground their ethical opinions in scientific fact. So assuming there is something rightly called ethical wisdom, where does it come from? There is considerable confusion about this, even among ethicists.

One virtue of perennial philosophy is that it has an answer to these questions. According to perennial philosophy, what is right and good is what conforms to divine knowledge of the right and the good. God is all-good and all-loving. He is the way, the truth, and the light. We are created in God's image and we have the capacities to see, know, or discover the right and the good. If we honestly search our hearts and minds, this thinking goes, we can see what is morally good. And then we can follow the moral truth so revealed.

What we philosophers call meta-ethics—the analysis of what moral terms mean and refer to—raises serious questions for ethics. The task of meta-ethics is to explain our received ethical wisdom in secular terms that preserve much of the wisdom of the ages. What Alasdair MacIntyre calls the "enlightenment project" was devoted to this task. John Stuart Mill and Immanuel Kant each proposed a secular ethical theory that yields an ethic very close to the one proposed by Jesus of Nazareth. But moral skeptics inspired by the scientific image immediately hit on the following worry. The world as science conceives it is composed of physical objects, events, and processes—things with mass, weight, texture, and energy. Saying "This bowl weighs 10 pounds" makes sense, as does saying "He is dead." But when we say "This bowl is beautiful" or "It is bad that he is dead"—where is the beauty or badness?

From a certain perspective, the answer to such questions is that beauty and badness are nowhere in the world-out-there. Either they are fictions akin to unicorns or mermaids, complete products of imagination, or more likely they are mental projections, most likely emotional projections. Here, scientifically inspired philosophy took a page from the perennial philosophy and accepted as a first principle that emotions are notoriously unreliable guides to truth. Emotivism holds that aesthetic and ethical judgments are projections of feelings. Thus they are not judgments based on reason and evidence, but rather akin to rooting for one's favorite sports team. Fun perhaps, done with conviction and all that, but not cognitively respectable. The reason one likes the team one likes is typically utterly arbitrary: One grew up in a certain locale, perhaps, among people who rooted for the same team. There is no way to claim that reason favors rooting for this team over that one, that this—among all the teams—is the *right* team to root for. Once ethics is conceived along emotivist lines, and once emotions are accepted to be nonrational, the idea that there is moral truth is in serious trouble.

The problem is this: Perennial philosophy says that there is an objective moral truth, a right way to think and behave, a divinely or rationally communicated pathway to attain goodness. But one variety of scientifically informed philosophy says that people cannot see the supposedly "objective" objects, properties, events, or processes that terms like "good" and "bad" and "right" and "wrong" refer to. If these words don't refer to objective things, they must express subjective states of mind. To say "It is bad that he is dead" might mean "I am sad that he is dead" or "I don't approve of the way he died." But though everyone will agree about the objective facts—he *is* dead—I, unlike you, may be glad he is dead and may think he deserved to die the way he did, say, at the hands of my brave soldiers. We have different feelings; we disagree. End of story.

The scientific image does not require the secular moral philosophy that pays it homage to take the emotivist line about ethics. But it has been hard to develop a conception of ethical knowledge that gives ethical judgments something resembling rationality and objectivity, once the physical sciences are taken as setting the standards for objectivity and rational belief.

Still, if we put our minds to it, it is possible to conceive of ethics as cognitively respectable. First, one can question the privileging of "the objective." Most of us accept that certain states of affairs are objective in the sense that they would be as they are even if no sentient beings existed to detect them. Physics tries to describe such a world, the world-as-it-is independent of any observers. We plausibly believe that the Newtonian law force = mass × acceleration would obtain even if we should all die tomorrow. But colors and smells and tastes would die with us. These things are not objective in the sense that they would be there if sentient beings weren't. So what? Surely, while we are here, there are colors and smells and tastes. As long as there are sentient beings it is an objective fact about the world that there are colors and smells and tastes. Furthermore, it isn't as if the colors, smells, and tastes of things are subjective in a way that makes my saying "That is red" or "This is sweet" fairly comparable to "The University of North Carolina's basketball team sucks." Colors, smells, and tastes require that there be subjects of experience. Once there are such subjects then there is all sorts of consistency in how, within a species, these things are experienced.

Second, one can question the premise that all emotions are nonrational, fickle, and flighty. Sensation and perception are typically reliable and emotions are normally apt. Is it nonrational to fear great bodily harm? Is it nonrational to be angered if someone threatens my loved ones? Are love and friendship flighty and fickle? It seems not. So why paint all emotional reactions with a brush suited only for some emotional displays?

Third, science does employ "ought" talk: "If you want to get water and humans and livestock over this river, then you ought to build an aqueduct and viaduct. Here are some designs that will work."

In technological applications, science takes certain ends as reasonable and specifies the means to attain them. Is there a difference between ought talk in science and ought talk in ethics? I don't see that there is. In both cases, we discuss and then specify certain reasonable goals, aims, or ends, and we consider the best means to achieve them. People wish to flourish. Wisdom teaches that virtue and goodness, love and friendship, concern for those less fortunate, will help one do so. Ethics, as I see it, is part of human ecology. Ecology is a normative science: It studies how different life-forms flourish in their environments. Certain environments are objectively better for the flourishing of wetlands, beavers, orchids, and pine forests. Ethics is the normative science that studies the objective conditions that lead to flourishing of persons.

I'll spell out how to conceive of ethics' functions as a part of human ecology in the last chapter, but for now I'll offer this capsule summary. If ethics is conceived along the lines of human ecology, then it must be accepted that emotions are deeply involved in moral life; at the same time, this scenario denies that all emotions are nonrational, fickle, and flighty. Humans do better, indeed they flourish, if they feel, think, and act in some ways rather than others. One way that is verified involves the Golden Rule. The Golden Rule expresses ethical wisdom not because Jesus said it, but because he spoke truthfully when he claimed that following the Golden Rule is good, that it makes us flourish.

Mind, Morals, and the Meaning of Life

Here then is how I see things. There is a conflict between perennial philosophy's humanistic image and the scientific image of persons. The perennial philosophy says we are ensouled, the scientific image says we are not. The perennial philosophy says we are only part animal, the scientific image says we are all animal. These might seem like high enough stakes to justify the conflict and explain the impasse.

But the question What is a person? is a high-stakes game precisely because it has implications for the weighty questions of meaning and morals. Mind, morals, and the meaning of life are the ultimate stakes.

If the scientific image of persons and minds is to replace the humanistic image it will need to win and hold converts by demonstrating how it preserves some of the perennial philosophy's vision of what minds do. But this will not be enough unless the scientific image also shows that it has the resources to explain how life can have meaning and how morals are possible and rational. It can do this, I think, but little effort has been expended toward this goal. A central aim of this book is to show how to conceive of mind, morals, and the meaning of life from the perspective of the scientific image.

WE NEED TO REVISE the commonly accepted story of human nature. The reason to rewrite the story is that the current humanistic one is false. The time is ripe to come to grips with the image of human being that is available, in a nascent way, through a philosophy that is attuned to our best science. We don't yet see ourselves within the framework of such a philosophy, but with some work we can begin to construct a truer picture of ourselves.

Dasein and *Dharma*

What does it mean to be a person and how do we find our way? I have already introduced the term *dasein,* a German word meaning literally "there" (*da*) "being" (*sein*), which acquired a technical philosophical meaning in Heidegger's work. Heidegger was hunting for a neutral word in the vicinity of *human* or *person* or *the self.* He thought the word *human* was too closely aligned with the thin concept of a mere "member of the species *Homo sapiens,*" whereas *person* carried various, sometimes inconsistent, connotations—a mask (*persona*), one's true self, an admixture of mind and body, a responsible citizen, and the like—that depended too heavily on a particular philosophical or cultural conception. So *dasein*—"there-being"—was his choice. In addition to its promise as a neutral term from which proponents of perennial philosophy and the scientific image can debate the nature of persons, *dasein* offers the advantage of having arisen within the existentialist tradition of philosophy, a tradition grounded in an almost obsessive concern not only with what kind of being a person is but what, if anything, human life means and how we are to find our way, assuming we can, to fulfillment. To raise questions about *dasein* is to raise questions at one and the same time about mind, morals, and the meaning of life. Given that a scientifically inspired philosophy must be concerned with all three of these subjects, *dasein* is a good word to use.

Dharma, meanwhile, is a Sanskrit word with no precise English equivalent. Its original meaning has to do with sitting still, where sitting still and focusing on your own mental states can help you move down the right path and get you to where you should want to be on "The Way." The term finds its home primarily within the Hindu and Buddhist traditions. In both, *dharma* refers to practices that lead to enlightenment, happiness, and fulfillment. It has similar meaning to *Tao* in Taoism, which is usually translated as "The Way"—that is to enlightenment, wisdom, and goodness. Many *dharma* practices involve meditation (thus the etymological root in the act of "sitting still") whose purpose is self-awareness, discernment of what is true and good, dispelling false beliefs, and developing wholesome ethical attitudes. The aim is the wisdom that comes through enlightenment.

Dharma usefully expresses in a single word the ideas of wisdom and enlightenment. Dharmic practices aim to produce the sort of wisdom that results from seeing things in their true light. Dharmic practices are a way to "The Way." I hasten to add that nothing at all rests on understanding *dharma* in the way(s) Buddhists or Hindus do, although doing so will not hinder you—and it may help you—in the project of seeking enlightenment.

There is another reason I find *dharma* useful. Most of this book argues that many of the philosophical beliefs that are most dear to our hearts—about the nature of mind, the self, free will, God, and the source of moral value—are false. These beliefs, though perhaps inspired by the love of wisdom, do not express, capture, or reveal true wisdom. *Dharma* means wisdom and enlightenment—things we seek, but currently lack when it comes to understanding our nature.

We do not see things, most importantly our own natures, in a clear, undistorted light. I do not mean to suggest that we have gotten everything totally wrong. We catch glimpses of mind, selves, agency, and goodness. But we see these things as they are reflected in poor lighting. Like the prisoners in Plato's cave, we mistake shadows and distorted images of the mind, the self, free will, and goodness for the real things. We mistake Appearance for Reality.

Using these shadows and images as the initial data, the "wisdom of the ages" has worked to produce elaborate theories of human nature. Like all good theorizing, it has deemed some evidence, some data, better than others—more pure, less distorted, comprising a more representative sample. This complex tapestry of theories constitutes our humanistic self-image. Like any carefully produced tapestry, it is strong and resilient, and its various parts produce a coordinated, beautifully patterned whole.

Imagine that there are many replicas and variations on the "Mother Tapestry" adorning the walls of our homes and museums. We see ourselves, possibly different aspects of ourselves, our lives, our history, and our values in these offspring tapestries. Indeed, we are able to produce the offspring tapestries and recognize them as representations of our way of being because they are external representations of an image or images we already carry internally. Our tapestries remind us of who we are and reinforce the way we already see ourselves.

Since we are heirs to the image preserved most fully in the "Mother Tapestry" (but also represented in the many offspring tapestries), we see ourselves and our place in the world in accordance with the patterns it projects. In the complex tapestries that adorn our walls, we see a representation of ourselves. We are confident, even cocky, that we thereby see our nature as it really is and that our tapestries express both the way we are and the noble ideals we aspire to.

My claim is that when we honestly examine our most important beliefs about ourselves and our place in the world, as well as our ideals, aspirations, and views about what gives life meaning, we will discover that many of our most cherished beliefs turn out to be false and many of our ideals are less sublime than we think.[3]

---

[3]One way in which the falsehood of the humanistic image and the perennial philosophy that supports it causes trouble is related to its underlying individualism. The picture that is endorsed sees each human as housing an individual soul and sees his or her fate as turning on whether he or she lives well or badly. The aim of the game is personal, that is, to achieve individual immortality. Despite widespread lip service to the ideal of the Golden Rule, most people do not show very generous love or compassion. Thomas Hobbes, the great advocate of psychological and ethical egoism, put the Golden Rule this way: Do *not* that to another that which you would *not* want done to you. Jesus's formulation encouraged

Few of these beliefs are explicitly defended by professional philosophers. Some—the belief in a nonphysical mind is a good example—they explicitly reject. But these beliefs are held by most intelligent folk, including many very well educated individuals. They are also held, even if only implicitly and tenuously, by many professional philosophers. How is it possible for professional philosophers to reject some of the problematic beliefs but still be under their spell?

There are not two camps of philosophers on either side of some great intellectual divide, fighting ably over contested turf. The situation is more subtle and disturbing. Philosophy, in our time, suffers from the same sort of specialization and intellectual division of labor one sees across all disciplines. Philosophers are not encouraged to work on the Big Picture. Some do, and these individuals are the ones who see that the Big Picture is built on less than stable foundations, in some cases on sand. Others work ably on "their" problem. Some devote a career to interpreting the writings of a great philosopher such as Plato, Descartes, or Kant. Others work on issues within ethics, the philosophy of science, the philosophy of language, logic, epistemology, or metaphysics. There is little encouragement to develop a theory about how things hang together in the broadest sense. This explains why most philosophers don't have such a theory, and why they can be comfortable with the dominant way of conceiving things even if their work undermines one aspect of this received view.

Take the nature of mind. Most philosophers of mind reject the Cartesian view that humans are possessed of two parts, a nonphysical mind and a body. But many of these philosophers will claim agnosticism when asked how their rejection of mind–body dualism affects other ideas found in the same nest: the nature of the self, free will, personal immortality, the source of moral value, and the existence of God. They are interested in these questions, but they don't "work on

---

active love: Do unto others as you would have them do unto you. I don't think it unfair to say that most Westerners conceive of their moral duty as not to do harm, rather than actively doing good. Although the individualism of perennial philosophy is hardly the only cause of moral selfishness, I do think it is part of the reason. The picture that emphasizes that we are dependent rational animals, and that this life is the only life—our single chance to bring about good—might I think dislodge some our individualistic ways of thinking and acting.

them." The few who do work on all of these problems think the whole nest of ideas is untenable.

In any case, something odd and unfortunate has occurred in the practice of philosophy when the beliefs of professional philosophers, either those who take on the Big Picture or those who work on specific aspects of it, are so different from the philosophical beliefs of ordinary people. But that is the case.

False beliefs are the philosopher's enemy. To be sure, some false beliefs provide comfort in the short term, possibly over the life of an individual, even over many generations. This is one reason for their resiliency. But there is abundant evidence that false beliefs ultimately do more harm than good, and that people who overcome such beliefs can flourish, in part because overcoming what is false and illusory is liberating.[4]

The duty to expose false philosophical beliefs is not to be undertaken solely in obscure philosophical journals on the way to gaining tenure. Nor is it sufficient to spread the word only to the lucky few who take philosophy courses at universities. More than a few philosophers think philosophical problems are inherently unsolvable and that the purpose of philosophical training is to help students acquire the analytical skills to think clearly and profitably when faced with solvable problems outside of academia—on Wall Street, in law practice, or in the public sector and politics. But many other philosophers think that certain widely held "answers" to philosophical questions are transparent nonsense. But they expend almost no passion in teaching students to see these beliefs for the nonsense that they are. It is not clear why. One diagnosis, suggested above, involves the division of

[4]There is interesting work in social psychology on "positive" illusions. For example, most people, save for moderately but not severely depressed people, underestimate the probability, even when they are given base rate information about the population to which they belong, about the likelihood of their getting cancer, being in a car accident, getting divorced, and so on. This makes them feel good along life's way. It is an interesting question whether people really believe, as opposed to hope, that bad things won't befall them. In any case, the sort of false beliefs that I claim are the philosopher's enemy are distinctively philosophical, and they involve matters about the nature of mind, morals, and the meaning of life that all parties involved in the discussion and debate at least verbally agree should be seen as they truly are.

intellectual labor. Another, less kind diagnosis imputes paternalism and cowardice. On this latter diagnosis, these teachers are like Dostoyevsky's suffering cardinal, the Grand Inquisitor, in *The Brothers Karamazov*. The cardinal and a small group of fellow clergy are, in fact, atheists. They do not believe in God, but they pretend to because their charges, they think, need this belief.[5]

I am not sure to what extent such hypocritical paternalism explains why some professional philosophers treat their students' unjustifiable preconceptions with a gentleness the students surely deserve, but that their beliefs often don't. Whatever its root cause, there is within the academy a professional reticence about undermining such beliefs. Dostoyevsky imputes deep psychological reasons to the cardinal and his circle that justify their not only letting certain illusions take hold of the hearts of their charges, but actively working to inculcate these beliefs. I don't see my professional colleagues actively fostering belief in what they judge false. But we may suffer, on some level, from the delusion that our charges need beliefs that we ourselves think are misguided.

Another reason is a commitment to fairness in presenting both sides of a complex issue, combined with the liberal view that it is up to each individual to make up her mind about the issues as she sees fit. But this stance carries a certain negligence. If it is not good to believe what is unreasonable or false, we do our students no favor in allowing them to think they are using their minds well if they choose to believe what is false or unreasonable.

The aim of philosophy is, or at least once was, to locate wisdom when and where it could be found. The Stoics, the Epicureans, Socrates, Plato, and Aristotle all framed their projects in terms of seeking and teaching wisdom. Dharmic practices in the East still seek wisdom, even if most professional philosophers in the West don't.

---

[5]Marx famously thought that religion was "the opiate of the masses." Religion is a drug in the sense that it dulls the pain of a miserable life by promising a better one. In this way, religion gives false hope and keeps the masses from rising up and demanding a better life now. There is no doubt that Marx was onto something important. And thus it seems fair to say that certain illusions are positive. But they are only positive in certain objectively bad social situations. If these conditions can be changed, then what is "positive" in the illusion loses its rationale.

Jean François Revel, a distinguished French political philosopher and journalist, writes, "Nowadays in our scientific age, philosophers have abandoned the ideal of wisdom."

In the entry on "wisdom" in *The Oxford Companion to Philosophy,* John Kekes provides this definition:

> A form of understanding that unites a reflective and a practical concern. The aim of the attitude is to understand the fundamental nature of reality and its significance for living a good life. . . . [A]lthough wisdom is what philosophy is meant to be a love of, little attention has been paid to this essential component of good lives in post-classical Western philosophy. It is perhaps for this reason that those in search of it often turn to the obscurities of oriental religions for enlightenment.

Because I think philosophy is—or should be—in the wisdom business, a business that succeeds in part by exposing false beliefs, I think of my style of doing philosophy as a kind of *dharma* practice. I also have more than a passing interest in Tibetan Buddhism. I can't tell if Kekes's comment about people turning to "the obscurities of oriental religions for enlightenment" is intended as a lament or not. It depends, I suppose, on whether he thinks "the obscurities" of oriental traditions are genuine or only apparent.[6] If one comes to the study of Buddhism firmly in the grip of the dominant Western humanistic image, it will surely seem alien and obscure. But if you come, as I did, without being fully gripped by that image, you will find that Tibetan Buddhism provides a way of thinking about human nature that is more

[6]There is a serious problem caused in some measure by philosophy's neglect of wisdom. Look at the *New York Times* Best-Seller List and you will find all manner of "New Age" books. One very popular one, as I write, is Gary Zukav's *The Seat of the Soul.* Oprah loves the book. And there are portions of it that I like, that contain good advice, and that are evocative and provocative. But the metaphysics of the book—that we have souls as well as personalities, that we are continuously reincarnated (I'm only mentioning the mild stuff)—is philosophically irresponsible. And Zukav, as far as I can tell, feels no responsibility to offer any arguments for his views. One reason (I'm sure there are many) why such books have great popularity is that analytic philosophy, because it "religiously" avoids the wisdom business, has little to say that ordinary, intelligent people might find inspiring.

congenial than our own to the view suggested by contemporary mind science. In this book I have no intention of asking the reader to become a Buddhist. Furthermore, I think that the compatibility (and it is limited) between the Buddhist humanistic image and that suggested by contemporary mind science is, as we say, a pure cosmic coincidence. The Buddha was a master of introspection unmatched by any Western phenomenologist, but he knew little science.

That said, I will remark, throughout this book, when and where I see convergences between certain aspects of the Buddhist humanistic image and the one I am recommending. Beyond my personal interest in *dharma,* I have two reasons for doing this. First, as I have said, people have an understandable resistance to changing how they see things. For some, it is threatening to radically adjust their beliefs of what it means to be a person. This difficulty can be ameliorated by seeing others who see themselves in the recommended way and are able not only to go on, but to flourish in life. Second, many people think there is something mandatory and inevitable about the humanistic image they embrace. In my experience, this sense of mandatory belief is explained best by a conviction that one has seen The Truth. My own analysis is deflationary. Most people have lived sheltered lives and are, largely for that reason, parochial and chauvinistic in their attitudes. They think that what they believe is true primarily because most of those around them also believe it. This is baseless as well as irresponsible.

Thus the two main questions addressed here arise in the words *dasein*—Who are we?—and *dharma*—What is the way to meaning and goodness? The goal of this book is to seek wisdom and to expose false or untenable beliefs along the way. The false or untenable beliefs I have in sight are beliefs about ourselves—about our minds, the self, free will—and about human nature, goodness, and life's meaning.

How and what we are, and why we are as we are, represent profound questions that we would like to understand better. Contemporary mind science reveals much that is important about *dasein* and *dharma*—about our nature and the road to wisdom and enlightenment. It is thus to mind science that I turn for help in the project of coming to see ourselves more authentically, as we really are. The dominant humanistic image needs to be replaced, and there is a better one waiting on the horizon.

TWO

The Human Image

We firmly believe and profess without qualification that there is only one true God—Creator of all things visible and invisible, spiritual and corporeal. By His almighty power from the very beginning of time, He has created both orders of creatures in the same way out of nothing, the spiritual or angelic world and the corporeal or visible universe. And afterwards He formed the creature man, who in a way belongs to both orders, as he is composed of spirit and body.

—*Fourth Lateran Council, 1215;*
Reaffirmed by Pope Pius XII,
Humani Generis, *1950*[1]

Darwin's idea bear[s] an unmistakable likeness to universal acid: it eats through just about every traditional concept, and leaves in its wake a revolutionized world-view, with most of the old landmarks still recognizable, but transformed in fundamental ways. Darwin's idea had been born as an answer to questions in biology, but it threatened to leak out, offering answers—welcome or not—to questions in cosmology (going in one direction) and psychology (going in the other direction).

—*Daniel Dennett,* **Darwin's Dangerous Idea**, *1995*

[1]Pope Pius XII returned to the subject of human origins seven centuries after the Fourth Lateran Council because of the challenge of evolutionary theory. Pius conceded that God might have created Adam and Eve through a process akin to evolution. But Adam and Eve were not animals (the Pope insisted that all humans were descended from the primal couple) since God gave them souls, as He continues to do for all humans at the moment of conception.

EVERY GROUP OF PEOPLE, in its own place and time, has a commonly held picture of what a person is, how a good life is to be lived, and where meaning and purpose lie. This picture antedates new scientific discoveries, new scientific theories, as well as the public activity of scientists directly challenging entrenched views they take to be superstitious or misguided. When science, therefore, speaks on a matter that is already centrally situated in the dominant picture of persons, it directly challenges that picture. In this way it question the self-image of an age and by implication the image of the individual persons who see themselves in the terms set by the dominant picture.

Sigmund Freud's *The Future of an Illusion,* B. F. Skinner's *Beyond Freedom and Dignity,* E. O. Wilson's *Sociobiology,* Richard Dawkins's *The Selfish Gene,* Daniel Dennett's *Darwin's Dangerous Idea,* and Frances Crick's *The Astonishing Hypothesis* are all popular books that clearly try to get in your face. They do not merely suggest or hint at new ways of seeing ourselves, leaving their challenge to the self-image of an age implicit, leaving to the reader the work of accepting, rejecting, or absorbing the proposed view about religious belief—that it is based on an illusion, the falsehood of the idea of free will, the fact that genes hold culture on a short leash, that there is no such thing as a soul, and that meaning and purpose, if they can be found at all, must be found in terms consistent with the spirit of the neo-Darwinian theory of evolution. The number of such "in your face" books, coming from such distinguished figures, constitutes a major challenge to perennial philosophy. These writers say wake up and smell the truth—a truth that, *if* true, is guaranteed to disquiet, destabilize, possibly even explode the dominant image of our age.

In this chapter I continue, this time in a more autobiographical vein, to lay out the general problem of competing images, of which the competition between the humanistic and scientific images of persons is an instance. Then in the next five chapters, I'll speak directly about the implications that recent work in the mind sciences and in naturalistic philosophy has on our way of conceiving of the human mind, human agency, the self, and the good life.

Image Is Everything

There is an ad for Canon cameras that announces, "Image is everything."[2] Canon is not trying to provide philosophical wisdom or reveal deep truth. In the fall of 1999 The Gap informed all young people that they had better purchase a colorful fleece vest if they wanted to run with the in-crowd. Their ads said, "Everyone in vests." More recently, in preparation for the winter of 2000, they announced, "Everyone in leather." The Gap was not hawking wisdom either.

But there is a pearl worth examining, actually two pearls, in these ads. The first has to do with the idea of image, the second with the idea of an in-crowd.

First, there are several things that the motto "Image is everything" resonates with: that appearances matter, that how we are seen by others is important, that feeling good about oneself is tied to both how one sees oneself and how one is perceived by others. Ideally, the image one projects is a reasonable facsimile of one's self-image.

Second, with respect to the idea of an in-crowd, we are social animals, and as such belong to groups with particular values, interests, and beliefs—including beliefs about how to dress. Next spring, The Gap might try to sell togas. "EVERYONE IN TOGAS," the ads would implore, which might not go over that well. I might buy one, however, since one in-group I am part of is the group of philosophers—and Socrates, one of our idols, wore a toga.

In any case, we humans, at least those of us who exist now—and probably all who have ever existed—do not belong to one homogeneous group that shares all values and beliefs but to heterogeneous groups for whom the range of normal beliefs and values are defined by the group, the in-crowd, to which one belongs.

Putting the two points—about image and in-crowds—together, we can say that one's sense of self, of purpose, of identity is tied to inhabiting some space in the range of the normal for the group or

[2]There is a Sprite ad that announces "Image is nothing, thirst is everything." People do sometimes say that "image is everything." If they didn't the Sprite ad wouldn't work.

groups to which one belongs. We might then say, exaggerating somewhat, that self-image is everything, where self-image is understood as the way one sees oneself, indeed the way one needs to see oneself and one's place in the world, to feel good about oneself, to feel whole, to feel self-respect, to feel, as it were, like a person.

Self-Authorship

What I have said so far might make you think that one's sense of identity is completely determined by the group to which one belongs. This is not entirely false insofar as we are born into families and communities with an image of persons already in place. We have no say about the location in the space of images into which we are born. The image antedates us, often by centuries. At least at first, we have almost no authorial control over the story in which our own life is embedded and from whose perspective it is shaped. Once we reach an age where we do have some control, we work from the image, from the story that is already deeply absorbed, a story line that is already part of our self-image.

I was raised a Roman Catholic. When I was a five-year-old first-grader, we prepared for first communion and before that for our first confession. I was taught about heaven and hell and told that only Catholics could be saved. In preparation for confession we were told that we should confess such things as lying and profane language. I understood what these things were, even why they might be sins. My interest was peaked on the sin front when the nuns explained that we must also confess impure thoughts. At the time, I was clueless as to what an impure thought was and thus very curious. I harbored aspirations to one day have impure thoughts, whatever they were. The nuns clearly thought they were very important and very bad, so I remember confessing to six impure thoughts at my first confession. I then added one to the number of lies I confessed to, to cover this latest one.

When I was six or seven, I started lapsing. Mom had explained that in heaven you get to do whatever you want, which for me meant eating peanut butter and jelly sandwiches and drinking chocolate ice cream sodas all day long. Heaven sounded good, especially when contrasted with the option of burning in hot oil for all eternity. But by second grade I started to see pictures of God sitting on a cloud with the heavenly host appearing to spend their time singing hymns—in Latin I assumed—day in and day out, for all eternity. This seemed like an exceedingly boring prospect. I began wishing there were a third choice—of just dying and being dead for good. I didn't see why this wasn't preferable to singing hymns in Latin for all eternity or burning in hot oil for all eternity. I thought a good God would provide such an option.

But the main reason for my lapsing was that I came not to like or respect the God I was taught to believe in. I still don't like or respect the image of *that* God. Of course, I struggled considerably in my lapse and didn't really complete the process until I was 13 or 14, when a testosterone surge both gave me the courage to reject my faith and secured my belief that God was either mean or confused, possibly both, if he thought that sex was as bad as I had been taught.

How was I able to reject such a central theme of the story that set out the basic structure of my own life story—the story in which I was to find meaning and purpose, in whose terms I was to become a person? The answer is that there were alternative stories, alternative ways of living and believing coexisting in the neighborhood where I was raised. My beloved Uncle Austin was a lapsed Catholic. He was also very good and very smart. I got into a fair amount of trouble with my father when I overheard my mother telling Dad that Father O'Connor, our parish priest, had told her that she could not attend Uncle Austin's wedding because he was marrying an Episcopalian in an Episcopal Church. I yelled from the kitchen that Father O'Connor was a horse's ass—a locution I had learned at Dad's knees but that, in the present situation, did not go over well. But I got a sense that Dad, despite grounding me for my disrespect and profanity, agreed. So Uncle Austin, Dad, and Father O'Connor, each in his own way, abetted my lapsing.

Furthermore, my Jewish friends seemed to go light on the God stuff. They also thought that sex wasn't so bad. Some of them thought it was good, even fun.

I was able, then, to reject a central theme of my native Catholicism because there were other in-groups, groups of decent lapsed Catholics and decent Jews, in whom I saw alternative options. I worked to draw certain parts of their story into my own, without abandoning all the themes—many of which I powerfully identified with—of the story of the form of life within which my own life was to be situated. I doubt I could have found a way to reject my Catholic God had I not seen around me examples that revealed how I might pull off lapsing without losing myself altogether.

When I am introduced to people as Owen Flanagan, they often say, "Oh, Irish Catholic." I almost always say "yes," even though I have a lot of French Canadian blood and I am, as I said, lapsed. Being Irish Catholic is not only a way others situate me in the world; it is also, to this day, a way I situate myself. There is no other way for me to understand myself, for in many ways I see the world within terms set by the postwar American-Irish Catholicism in which I was launched and with which I still identify.

We can, in short, take our story in other directions, but usually only if we have available to us alternative stories, alternative plot lines, as live options—options that can be fitted somehow into the story already in place. America being a very pluralistic culture, there are a fair number of places to find such options. But conservatism reigns. You will be, as we say, messed up if you try to remake yourself with wholly new cloth.

Conceptual Schemes

Philosophers of science often speak of the idea of a conceptual scheme. Suppose you are a follower of Aristotle and Ptolemy and believe the earth is the stationary center of the universe. Your conceptual scheme is Ptolemaic. Copernicus and Galileo come along and propose an alternative: The earth is in motion and it orbits the sun, which is the center of our solar system but not the center of the whole universe.

For the Ptolemaic astronomer, the Copernican alternative is destabilizing, for it says that what the Ptolemaic believes is false. Normally, no respectable scientist simply abandons his theory when even a promising alternative presents itself. There is too much personal investment, and investment in the scientific in-group to which one belongs, to simply switch allegiance.

When I read that eggs are bad for me I cut back my egg consumption. Years later I read that doctors think eggs are okay, even good for you, so now I eat more eggs. I find these changes in medical wisdom annoying, but despite requiring slight adjustments to my conceptual scheme, they are no big deal when it comes to my self-image.

Religious, ethical, and scientific conceptual schemes are different. They take up a lot of space in the minds of adherents, they sit at or close to the center of the conceptual scheme, and they are often woven together. Even small adjustments can have threatening implications.

Copernicus died in 1543, the year his *The Revolution of the Heavenly Orbs* was published. He therefore got into no trouble with the protectors of the self-image of his age. Galileo of course got into a great deal of trouble with the Catholic Church when he defended the Copernican view seven decades later. Why did the church care about astronomy? The simple—but pretty accurate—answer is that the Ptolemaic story fit neatly into a view of God's plan for creating humans in his image and giving earth pride of place in creation. The Galilean view destabilized this picture at a time when there were already too many heretics, thanks to the Protestant Reformation, running around Europe and causing no small amount of trouble for Roman Catholic orthodoxy.

Suppose that when the Copernican alternative was offered it was a matter completely internal to science, of no interest to defenders of the humanistic image. The battle would have been fought completely on grounds of empirical and theoretical adequacy and mathematical simplicity. And suppose that in this battle, the Copernicans, with help from Galileo and Kepler (who figured out that planetary orbits were elliptical, not circular), won the day, as they in fact eventually did. Now imagine a forlorn defender of the Ptolemaic view who has accepted defeat saying to his therapist, "We have lost the earth. There is no earth. Earth doesn't exist."

Is this guy out of touch with reality? Maybe not. There is a sense in which Ptolemaic astronomy and Copernican astronomy can be said not to mean the same thing by the term "earth" since they assign it radically different, indeed mutually inconsistent properties. According to the first view, the earth is stationary and the center of the universe. According to the second view, the earth is in double rotation—first, on its own axis, second, around the sun—and is not even the center of the solar system, let alone the universe. The therapist, if she is wise, will see what it is that legitimately worries the Ptolemaic. She might try to relieve his despair by saying, and saying truly, that there is still a sense in which the two views do mean the same thing by "earth," namely, "this heavenly body we call home." But the Ptolemaic may rightly respond that he meant much more by "earth" than this thin idea, this primitive core sense. By "earth" he meant this heavenly body we call home and that is stationary and lies at the center of the universe. Earth, in his sense, no longer exists. He sees this and even accepts that it is true, but it sickens his soul.

Keep this forlorn Ptolemaic in mind. His image will help us, in later chapters, to understand why some think that there is no such thing as "mind" or "the self" or "free will" if each is reconceptualized in the terms required by contemporary mind science.

THE IDEA OF A CONCEPTUAL SCHEME, of a connected set of beliefs that help constitute a worldview, is useful. But for my purposes, it is better to speak of a *conceptual-conative scheme,* for the kind of scheme I am interested in exploring is not just a connected set of beliefs but a set of beliefs and values to which a group and its members are deeply attached. Only if we understand these sets of beliefs, whether humanistic or scientific, as conceptual-conative schemes can we explain the forlorn Ptolemaic.

What I've been calling the humanistic image is a scheme that depicts the nature of persons, in whose terms one's self-image is defined and according to whose vision one tries to live one's life. A scientific image, despite possibly not speaking to the nature of persons, may well be essential in constituting identity for its adherents. The scientists' life work may, after all, be tied up with establishing the truth of some theory. This is why it is rare for defenders of well-worked-out scientific theories to concede defeat. The Ptolemaic who accepts the victory of the Copernicans is a rare bird. A new paradigm achieves complete victory over its predecessor only when the last of the old guard dies off. The old boys seldom concede defeat while they are alive.

The point is that different beliefs involve different degrees of personal investment. What I am calling a conceptual-conative scheme, Willard Van Orman Quine calls the "web of belief," and he compares it to a spider web (see Fig. 2.1). When the web's periphery is damaged, repair is simple. When the center is damaged, you hit me, the spider, where I live, and I may have to rebuild from scratch. I may need to move and build a new web elsewhere. Some kinds of damage to an existing conceptual-conative scheme may be easily fixed; other kinds of damage, depending on how close to the center they hit, are more costly. For Quine the laws of logic occupy the center. Disrupt them and I no longer know how to think or speak.

Good diet matters to me. I pay attention to what is said about vitamins and minerals, fiber, cholesterol, proper weight, and exercise. But when the medical establishment changes its views about proper nutrition, my worldview is hardly destabilized. When a fellow philosopher or economist says that we are all rational egoists playing some self-serving strategic game, I am more discomforted. I need to work hard to see whether and how this could be true, and if I see that it is true, I need to make some fairly deep adjustments in how I see myself and others.

FIGURE 2.1

When seventeenth-century Christian intellectuals saw an alternative to the received view of an earth-centered universe being floated by very wise scientists, the received view of persons and their place in the universe, and by implication the self-image of each person whose life was lived in terms of that image, was shaken up. Their conceptual-conative scheme told them humans are created in God's Image, that we and our home are at the center of creation, indeed that we are the *raison d'être* for creation. When such an image is challenged, so is the self-image of an age. When Darwinism explains that we are animals, descended from earlier species, the picture that says we are not animals, that we sit high on the Great Chain of Being, beneath God and angels but above animals, takes a hit. Not surprisingly, this is disquieting and destabilizing.

There are several responses: Show the disquieting and destabilizing alternative to be false on evidence that both sides are committed to accepting; dig in your heels and insist that the new view must be false because the image already in place deems it false; or adjust the conceptual-conative scheme to absorb the new view at minimal cost to its overall structure. With the Copernican and Darwinian revolutions, all three strategies have been tried.

The Manifest and Scientific Images

There is an old philosophical problem about appearance and reality: How things seem is not always how they are. The moon looks small, but it isn't. Tables seem solid, but they aren't. They are fields in Hilbert space—a quantum mechanical way of saying they are largely made up of empty space. I am pretty sure that the mind is the brain, although I admit it doesn't seem that way to many people.

In the mid-1960s, the philosopher Wilfred Sellars drew a distinction between what he calls the manifest and scientific images that captures the aspect of the appearance–reality distinction I am interested in: We humans *seem* to ourselves differently from the way we *really* are. Sellars draws the distinction this way:

> The "manifest" image of man-in-the-world can be characterized in two ways, which are supplementary rather than alternative. It

is, first, the framework in terms of which man came to be aware of himself as man-in-the-world. It is the framework in terms of which, to use an existentialist turn of phrase, man first encountered himself—which is, of course, when he came to be man. For it is no merely incidental feature of man that he has a conception of himself as man-in-the-world, just as it is obvious, on reflection, that "if man had a radically different conception of himself he would be a radically different kind of man."

He then adds:

I have characterized the manifest image of man-in-the-world as the framework in terms of which man encountered himself. And this, I believe, is a useful way of characterizing it. But it is also misleading, for it suggests that the contrast I am drawing between the manifest and the scientific images, is that between a pre-scientific, uncritical, naive conception of man-in-the-world, and a reflective, disciplined, critical—in short a scientific—conception. This is not at all what I have in mind. For what I mean by the manifest image is a refinement or sophistication of what might be called the "original" image; a refinement to a degree which makes it relevant to the contemporary intellectual scene.

What Sellars is getting at can be interpreted this way: There was a time—perhaps in the Late Pleistocene, tens of thousands of years ago—when humans *seemed* a certain way to themselves; the world they inhabited also *seemed* a certain way. I doubt that our early ancestors had a worked-out conception of how things seemed. Nonetheless, they themselves and the world they inhabited seemed a certain way to them. This original image was, for all intents and purposes, prescientific, uncritical, naive, and probably inchoate.

Time passed, and the manifest image became self-conscious and articulatable in folk wisdom—in stories, rituals, and art. I don't know when what we would call science began. But we do know that there were among the early Greeks a group known as atomists who postulated that solids were in fact made up of indivisible and unobservable particles. The most notable of these thinkers were Empedocles, Anaxagoras, and Democritus, all born in the fifth century B.C.E. And astronomy was old hat by 399 B.C.E., when Socrates was offered a hemlock cocktail on the house. As most everyone knows, Socrates was sentenced to death for corrupting youth by teaching them to question authority—of which he was certainly guilty. But one thread in the charge was that he corrupted the youth by doing science, in particular by teaching that heavenly bodies were rocks, not divinities. This thread ultimately played no role in his conviction, first because Socrates pointed out that he did not do astronomy, and second because he pointed out that the works that claim that the moon and its mates are rocks were readily available in bookstalls at the local marketplace, at *Agora.com*, as it were.

In any case, there came a time when attachment to a certain image took pride of place in the minds of a variety of groups. The manifest image, at one time prescientific, precritical, and naive, has not been so for at least three millennia. The Old Testament and the thinking of Socrates, Confucius, and Buddha, all of whom lived almost 2,500 years ago, present mature, critical images of humans and of what it means to live a good human life. Agriculture, economics, psychology, physics, botany, zoology, and astronomy existed and were absorbed in various ways. The science of Democritus, who first proposed the atomic theory of matter; of the Greek natural philosophers who thought that all of creation was made up of the four basic elements; Ptolemy's astronomy; and Plato's privileging of mathematics as a special type of knowledge, one that is hard to discover but crystal clear and certain once achieved, and that is to be emulated as far as possible as we seek to understand The Good—all of these threads had worked their ways into images that provided a way of seeing oneself, one's world, and one's place in that world.

Sellars uses the term *manifest image* to refer to the ordinary, commonsense image that takes as its subject both human nature and the nature of the external world. We can think of it as the composite set of all the folk theories of ordinary people. What I have been calling the humanistic image is the part of the manifest image that concerns the nature of persons. We might say that the manifest image has two components—the humanistic image and the world image. *Weltan-*

schauung, the "worldview" of German philosophy, has the same bipartite structure. The two components, of course, interact. Some Greek philosophers who embraced the atomic theory of matter, for instance, applied that theory to persons. These philosophers were the first card-carrying materialists. Many of their fellow philosophers did not think the atomists' theory of the material world, even if true, applied to persons. These folks were mind–body dualists.[3]

In any case, the humanistic image is a subset, an essential component of the manifest image, the composite set of all the folk theories of ordinary people. It is fairly easy to isolate the humanistic image from its manifest companions. The humanistic image is the part of the manifest image that takes human nature, human being, as its subject. And it can and often does display a sophistication due to absorption of theological, philosophical, and some scientific ideas.

One other point for the sake of clarity. I have said that the conflict between the manifest image of a people or an age and the scientific image is just an instance of the more general problem of clashing images. Scientific images can clash as well: Again witness the clash between Ptolemaic and Copernican astronomers. This clash had consequences for the humanistic image, because that image was opinionated about the lack of motion and central cosmic location of our home and had woven those opinions into its theory of the meaning of human life. If it had no bearing on these matters, the whole conflict would have been completely internal to science. If, on the other hand, and more plausibly, people maintained the seemingly (still) perfectly sensible view that the earth doesn't move but did not assign any larger meaning to this fact, the matter would have been of consequence only to the part of the manifest image that held opinions about the external world. If the Copernicans then won out (as they did), their victory would not require adjustments to views that pertain to human life, save indirectly by reminding people once again that they can be mistaken.

[3]The great physicist Niels Bohr once said that today's science is tomorrow's common sense. This is sometimes true, but not always. If it were always true, Darwinian evolutionary theory would not be so strongly resisted in many quarters nearly a century and a half after Darwin published *On the Origin of Species*.

Just as scientific clashes may remain completely internal to science, the humanistic components of manifest images can also clash without involving any conflicts over scientific matters. Athenians and Spartans had different views of the character traits and political practices by which humanity was best actualized; they could articulate (if incompletely) their competing manifest images, and these deep differences in their respective manifest images explain in large measure the mutual disrespect in which they held each other. Conflicts between Al Qaeda and the West, between Pakistan and India and among different ethic groups in Bosnia are clear contemporary cases of conflict bordering on revulsion, between different humanistic images. The strife between Israelis and Palestinians is another example that involves mutual hatred and suspicion, in part, because of seemingly irresolvable theological interpretations over the fate of Jerusalem. Science, of course, has nothing to say about whom God intended to reside in Jerusalem. But if pressed for an opinion, I think most scientists would say it is absurd to think that anyone knows what God intended for any piece of earth.

The distinction between the manifest image and the scientific image is not pure because, as I have said, scientific beliefs can be absorbed into the manifest image—sometimes effortlessly. Furthermore (although Sellars doesn't make this point), work in the human sciences invariably proceeds with some assumptions of the manifest image in place—in our time, for example—with the assumption that we are conscious creatures and that in some sense of the term we are agents, not mere automata.

Sellars insists, however, that "There is . . . one type of scientific reasoning which it [the manifest image], by stipulation, does *not* include, namely that which involves the postulation of imperceptible entities, and principles pertaining to them, to explain the behaviour of perceptible things."

This stipulation seems to me unnecessary and unhelpful. The atomic structure of matter, for instance, is part of the commonsense worldview of ordinary people. Sellars has told us he will not require the manifest image to be naive, to include only how things *seem* from some untutored, unacculturated perspective, but this move requires naiveté.

Although the world appears to us as filled with solid objects, the atomic theory of matter is now in our blood and bones. It is not a mistake therefore to say that solids *seem,* in some sense of the term, to be constituted by molecules, atoms, electrons, protons, and the suite of yet smaller components. We accept other imperceptibles as well: the ultraviolet rays that give us sunburn, the radio waves that carry the morning drive-time talk shows.

The moon looks small to me, but it doesn't, again in some sense of "seem," seem small to me. If this is right, then it is right because some parts of scientific understanding have been deeply internalized and now reside within my image of the world.

It seems to me that my mind is between my ears. But perhaps I think so because I have absorbed a certain theory about the location of my mental life. Ancient Egyptians, Mesopotamians, Jews, Hindus, Chinese, and Greeks all privileged the heart as the seat of the soul while occasionally also attributing mental functions to the spleen, bladder, liver, and head. I suspect, although I can't prove it, that mental life *seemed* to occur in these other sites; that is, it was not that these ancient peoples believed that mental life occurred in the heart in the way we believe that all mammals have periods of REM sleep. Mental life *seemed,* to the ancients, to take place in the heart.

The fact remains that once a manifest image is in place, science can challenge, disquiet, and destabilize. This can happen if those in power have an explicit opinion on the matter about which the relevant science speaks. The church was explicitly committed to the Ptolemaic view of planetary motion when Copernicus and Galileo came along.

Science can also challenge a manifest image that is not so explicitly committed. Nowadays, for instance, many of us are confused about what to think of human genetic manipulation. In part, this is because our humanistic picture is not explicitly opinionated about the matter. We didn't see the prospects of gene manipulation coming until very recently. Of course, we live in a pluralistic society with multiple manifest images, or (better) with component parts of the overall image that differ in sometimes significant ways.[4] Some of these images, the ones that say don't mess with God's creation, do have opinions about genetic manipulation, even if these were not explicitly formulated in advance. In such cases, we can say that the manifest image has clear implications for how it is to respond to a certain class of scientific discoveries, or to the technologies the discoveries make possible. Most people I know accept our basic understanding of genetics. But it follows from at least some religious conceptions of persons that you should oppose genetic tampering.

This shows how an image is more than a way of seeing things. It can also guide action. A good human life is lived according to the terms it sets. It would be too wordy, but more accurate, to call the sort of scheme I am speaking about a *conceptual-conative-action guiding* scheme.

There is a deep, albeit obvious, reason why the manifest and scientific images can come to war. Each wants to endorse its image as *real*, as *true*, as *the right way to see things*.

Sellars writes:

> Let me introduce another construct which I shall call—borrowing a term with a not unrelated meaning—the perennial philosophy of man-in-the-world. This construct, which is the "ideal type" around which philosophies in what might be called, in a suitably broad sense, the Platonic tradition cluster, is simply the manifest image endorsed as real, and its outline taken to be the large-scale map of reality to which science brings a needle-point of detail and an elaborate technique of map-reading.

He goes on to say:

> The perennial philosophy is analogous to what one gets when one looks through a stereoscope with one eye dominating. The

[4]This claim about plural manifest images might seem to undermine the claim that I can sensibly speak about "our humanistic image" in the West. Not quite. The point about plural images does remind us that we are speaking about some sort of abstract, ideal type. The judgment of whether two images, abided in the same cultural space, are different will depend on the level of grain one is examining. It will also depend on where (in the center or toward the periphery) of the web of belief the noted differences are situated. Christian Scientists do not believe in seeking medical care for disease. Most people think this foolish. Except for this big difference in the belief about the worth of medicine and the consequences it has for associated views about proper care of one's children and oneself, Christian Scientists share most of the views central to the prototypical American Christianity.

> manifest image dominates and mislocates the scientific image. But if the perennial philosophy of man-in-the-world is in this sense distorted, an important consequence lurks in the offing. For I have also implied that man is essentially that being which conceives of itself *in terms of the image which the perennial philosophy refines and endorses.* I seem, therefore, to be saying that man's conception of himself in the world does not easily accommodate the scientific image; that there is a genuine tension between them; that man is not the sort of thing he conceives himself to be; that his existence is in some measure built around error.[5]

Sellars does not immediately endorse this conclusion, in part because he understands that both the manifest and scientific images of an age are revisable. Both are in some measure false, "built around error." But he is also hopeful that the two images might in some age "blend together in a true stereoscopic view." I share his hope. But it is clearly not the way things are now. Many deeply religious people resist the framework of Darwinian evolution, not just because they believe in the origin story of Genesis but because they can't accept that we are just animals. And because they think humans have souls.

Most people appreciate astrophysics. It produces amazing facts and breathtaking pictures. But many people grow wary when astrophysicists speak about the origin of the universe. They don't much like the idea that the universe could be a surd, that the material available for the Big Bang could have just been there, or could have emerged from prior natural processes that were always there—in the way, for religious people, God was always there. These two cosmolo-

[5]Sellars adds: "The scientific image of man-in-the-world is, of course, as much an idealization as the manifest image—even more so, as it is still in the process of coming to be. It will be remembered that the contrast I have in mind is not that between an *unscientific* conception of man-in-the-world and a *scientific* one, but between that conception which limits itself to what correlational techniques can tell us about perceptible and introspectible events and that which postulates imperceptible objects and events for the purpose of explaining correlations among perceptibles. . . . Our contrast then, is between two ideal constructs: (a) the correlational and categorical refinement of the 'original image,' which refinement I am calling the manifest image; (b) the image derived from the fruits of postulational theory construction which I am calling the scientific image."

gies may be fundamentally and forever incompatible. Consider the following familiar images from Gestalt psychology (Figs. 2.2–2.4):

These perceptual cases show that some competing images cannot be blended in true stereoscopic vision. Maybe the hope of blending the scientific image with the manifest image is as idle a hope as being able to see both the duck and the rabbit or the vase and the face at once.

Or consider the case of the philosopher's table. Right now I am sitting at a small desk, this philosopher's table. The desk is solid. It holds my computer and stacks of papers. Sir Arthur Eddington, the great English physicist, describes sitting down to write his book, *The Nature of the Physical World*, at a table similar to mine, his solid desk. But he remarks: "I have drawn up my chairs to my two tables. Two tables? Yes; there are duplicates of every object about me—two tables, two chairs, two pens."

FIGURE 2.2

FIGURE 2.3

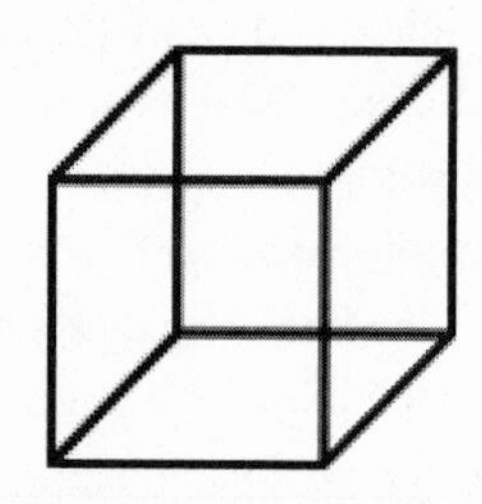

FIGURE 2.4

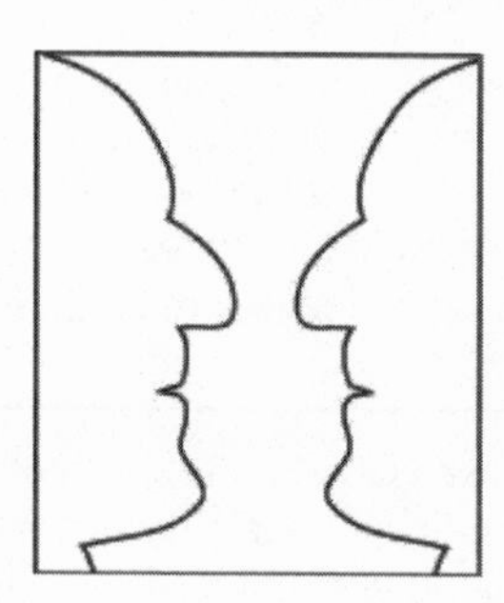

Eddington is not hallucinating. He means that there are two tables. One is the table as seen from the perspective of the manifest image, the solid everyday one; the other is the table as seen from the perspective of the scientific image. He writes of the second table:

> It is a more recent acquaintance and I do not feel so familiar with it. . . . My scientific table is mostly emptiness. Sparsely scattered in that emptiness are numerous electric charges rushing about with great speed; but their combined bulk amounts to less than a billionth of the bulk of the table itself.

Strictly speaking, Eddington sees only the solid table of common sense. But he knows enough about its underlying nature to know that it is in fact mostly empty space. The situation is puzzling.

Should we agree that there really are two tables? Or should we say that there is only one table and that its real nature is that revealed by science? Can we say that there are two perfectly good ways of describing tables, where "good" means good for two different purposes? Perhaps, talk of levels will help us achieve a consistent, coherent picture. The table is solid when viewed from the level of ordinary sensory perception and it is made up of mostly empty space when viewed from the level of particle physics. Can Eddington, or anyone for that matter, really hold in his head, at one time, both descriptions or must he shift back and forth between them? Whatever one's answer, one can see that Sellars's hope of blending the two images "in a true stereoscopic view" may not be possible.

The problem of free will and determinism has sometimes been described as Gestalt-like. Suppose I accept that everything that happens has a set of causes sufficient to produce it. Given this set of causes, whatever they are, this effect, whatever it is, is necessary. Suppose I also accept that when I act I do so for reasons, often after I think and deliberate. Such acts are free. But when I act freely, we can ask, why did I think, deliberate, and reason in the way I did rather than in some other way? Reasoning, thinking, and deliberating are causes of my act, but if causal determinism is true, these causes are themselves effects of other things. Indeed, my desires, my motives, my thoughts, my reasons, my style of deliberation must all have causes sufficient to produce them. These things seem to be under my control, to originate in me, but they aren't and don't. There is no denying that I reason, think, and deliberate; what is denied is that my reasoning, thinking, and deliberating are self-originating, things that cause my actions without themselves having causes that make them necessary. But if my thoughts are necessary then there is a problem with the idea that I could have acted other than I did. My feeling of choice comes from focusing on the proximate causes of action, the thinking, reasoning, and deliberating that occur just before I choose to act. If, however, I think of my action in terms of the wider causal nexus, as falling under the law of universal causal determinism, then I can't conceive of my action as having been different than it was. Each way of seeing my action—as free or as causally determined—can come clearly into view, but not both at the same time. The two pictures are Gestalt-like. Seeing my action as free makes seeing it as determined impossible, and seeing my action as determined makes seeing it as free impossible. Or so many think.

Some think this in addition: I cannot consistently think of my actions as causally determined. I must think of myself as a free agent. I must act *as if* I were free whether I am or not. This thesis is related to the first, but is somewhat different. The first claims that there are two incompatible ways of conceiving of human action and that when actions are seen one way they cannot simultaneously be seen the other way. The second thesis claims that the free way of seeing actions will normally be dominant. I will examine both theses in Chapters 3 and 4. For now, the point is that the problem of free will versus determinism has seemed to many to have the same structure as the Gestalt perceptual cases and therefore may be a case where blending the two competing images into one is impossible. This prospect is disturbing because unlike the perceptual cases, where there is no right or wrong answer, most people think the issue of free will versus determinism does have a correct answer. If causal determinism is true and if I do, or must, conceive of my actions as if they are free, then the conclusion seems inescapable that I am in the grip of an illusion, that I think of my will and my actions as being in a way that they are not.

Nonoverlapping Magisteria

In 1999, Stephen Jay Gould published the book *Rock of Ages: Science and Religion in the Fullness of Life,* in which he proposes a model for a solution to the apparent conflict between the manifest and scientific images. In our time, as in earlier ones, the conflict is between images supported by systems of religious belief versus those supported by scientific theory.[6] Gould's idea has received much attention in the popular press thanks to the debate in Kansas about teaching both the theory of evolution and the theory that the universe originated with the Big Bang. In 1999 the State Board of Education—amid much rhetoric to the effect that Darwinism rendered human existence meaningless—ruled that there would be no state testing that required knowledge of the Big Bang theory or Darwinian theory, effectively making such teaching unnecessary and possibly imprudent.

Gould's thesis is that religion and science belong to different, nonoverlapping and thus noncompeting magisteria, where a magisterium is "a domain where one form of teaching holds the tools for meaningful discourse and resolution." According to Gould, "Science covers the empirical realm: what the universe is made of (fact) and why does it work this way (theory). Religion extends over questions of ultimate meaning and moral value."

He then proceeds, using what he thinks is a settled matter in philosophy—the idea that there is a strict divide between "is's" and "oughts or norms"—to propose that science deals with the former and that religion and its accompanying manifest image deal with the latter. The two magisteria range over different things and do not overlap, and therefore do not—indeed cannot—conflict. Any appearances to the contrary are based on a confusion of the proper domain of

[6]This is one of those places where I will speak of the conflict between science and religion. But, I do not mean to introduce a third image, the religious image, in conflict with the humanistic and scientific images. Religious ideas play a powerful role—sometimes explicitly, sometimes implicitly—in the perennial philosophy that refines and endorses the manifest image.

each, or an overreaching of one domain into the territory of the other. This is a sweet but silly proposal.

First, Gould introduces religion in a way that makes belief in a God optional, almost incidental. Again and again, we are told that religion deals with questions of life's meaning and with ethical values. Theology is consistently absent from his description of religion's magisterium. Second, Gould fails to adequately recognize, let alone deal with, the fact that most of the great world religions see themselves as providing origin stories. That God is the creator, the causal origin of the cosmos, and the ground of meaning is at the core of Judaism, Christianity, and Islam orthodoxy. Unless these theological commitments are seen as myths awaiting a better scientific theory (which they are not) or as compatible with any future scientific discoveries (which they sometimes are, but not usually), then something is wrong with Gould's proposal.

Even if Buddhism is a major exception in this regard, Gould is not writing about a problem Buddhists have but about one that primarily afflicts American Christians—Christian fundamentalists in particular. Other strains of Christianity have less of a problem. Pope John Paul II teaches that evolution is compatible, or not obviously incompatible, with God being the creator.[7] French deism in the eighteenth century restricted God's role to the original creative act: putting the singularity in place that could then allow the universe to unfold in its own terms. No doubt religious authorities could also say that we do not while on earth have souls, but that God, being all-powerful, is able to reconstitute us as incorporeal beings after we die. But I am not aware of any religious effort to reconcile a picture of the world in which universal causal determinism reigns with one in which free will exists.

Gould is able to offer his principle of nonoverlapping magisteria because he is an agnostic and his understanding of religion is, to put it in familiar terms, that of the secular humanist. But the secular humanist is not religious in any familiar sense. He or she sees religious

[7]Whether the church's way of fitting evolution in is, in fact, compatible with the theory of evolution is dubious since Pius XII required, and John Paul II has not spoken against this idea, that in the human case God introduces souls into bodies at conception. God, of course, does not do this for bonobos, chimps, or gorillas.

origin myths as convenient stories that are placeholders for the yet-to-be discovered truth. And he or she sees resources to locate meaning and value in a godless world. Gould admits this much in several places when he points out that nonreligious philosophy—the kind that I and most of my colleagues do—has the resources to find sources of meaning and moral value in the world as conceived by science. Truly religious people don't see things this way.

The members of the Kansas Board of Education take religious origin stories to be true as origin stories, as correct theories about how the universe and humans were created, to which evolution and scientific cosmology are incompatible alternatives. Furthermore, they are undoubtedly troubled by certain scientists' claims that the latter theories are based on facts and that they render theological creation stories unnecessary, incoherent, or impossible—just as the epicycles of Ptolemaic astronomy became unnecessary with the victory of Copernican astronomy, or as the idea of a *nonphysical* human mind is said to be rendered incoherent by modern cognitive science and philosophy.

It may be that the question "Why is there something rather than nothing?" is one we can never answer. This does not mean of course that one is free to answer it any way one wants.

In any case, Gould's proposal misunderstands the magisterium of religion and gains its conclusion of nonoverlap and nonconflict on the cheap. His picture might one day become true. But it will become so only if the world's great religions reconceive their domain and range, or, what may amount to the same thing, if all religious persons become secular humanists. But this would be tantamount to their becoming nonreligious. Gould writes, "I believe, with all my heart, in a respectful, even loving concordat between science and religion." Sweet dreams, Professor.

Sophisticated Conciliationism

There are other advocates of something like Gould's principle of nonoverlapping magisteria, but who, unlike him, are not agnostics or secular humanists but theists, true believers. Ian Barbour and John Polkinghorne are major twentieth-century Christian theologians possessed of deep knowledge of contemporary science, especially physics. Nancey Murphy is a younger theologian who, more than these two, is up on the latest in mind science. These and other theologians are working to put science and religion on friendly terms. They emphasize that both are in the truth business, which in both cases is a historical quest that can never be finished—can never be, at any given time, at *the* point of having grasped the whole truth. For the scientific and theological images to be consistent, however, I believe theologians will have to concede more than they want, indeed will have to risk becoming nontheological and accept that talk of the divine involves overreaching.

But let me first say a bit about how this sophisticated conciliatory view works. First, there is the insistence that theology is not to be subservient to science. There is evidence for theological claims, and the quarks much loved by contemporary physicists are as invisible as God is. Second, if God is creator, the created world cannot be inconsistent with his nature. The work of the conciliationist is to make clear how the created world and God's nature are consistent, and this will involve modifying the claims of religious traditions that stop the clock, as it were, and speak with excessive confidence about what God's nature is, for example, that he created the world just as Genesis says he did.

Thus the sophisticated conciliationist, unlike Gould, does not shrink from placing God in his traditional role of creator. But like Gould, he will happily adjust his view of how God created.[8] This is all good. But it does not resolve the contemporary American debate about the Big Bang and evolution, since Christian fundamentalists are unwilling to adjust their view about how God created the world.

There is another problem with sophisticated conciliationism, although it has less to do with the conflict between science and religion than with the conflict among competing religious conceptions. Polkinghorne believes that no matter what science says, the job of the Christian theologian is to explain how Jesus Christ could be God incarnated. Even if this can be done in a way that science cannot

[8]With respect to evolution, for example, Murphy does not invoke the miracle that Pius XII and John Paul II insist takes place in the human case at conception, where God implants souls in each body.

claim is inconsistent with anything it says, there is still a problem with non-Christian traditions. Polkinghorne is clear that in addition to making itself consistent with science, Christianity also needs to make itself consistent with other religions. He thinks this will not be that hard to do with Judaism and Islam since there are important commonalities—especially, a commitment to monotheism. But there is *no* commonality when it comes to divinizing Jesus Christ, which Polkinghorne, at least, thinks is mandatory for any adequate religious conception. Furthermore, Polkinghorne frankly admits that he doesn't have a clue about how the conciliatory project will work with Buddhism and Hinduism, since the former wants nothing to do with talk of a creator God, and the latter is polytheistic.[9]

This problem of reconciliation is a serious problem for the theologian, but thankfully not for me. I mention it to point out how even very liberal religious epistemology differs from scientific epistemology. As I've said, the sophisticated conciliationist claims that both science and religion seek the truth, where the truth is cashed out not in terms of certainty but in terms of (we hope) increasing verisimilitude. Furthermore, given our finite natures and the complexity of the universe, we have no reason to expect that this project will gain an end point at absolute truth. Increasing verisimilitude is all we can reasonably expect of either science or theology.

All this seems plausible. But there is an asymmetry between the way the worldwide community of scientists works and the way the sophisticated conciliationist does. Not to put too fine a point on it, but science requires that it works to sort out competing claims as quickly as possible. When physicists in Copenhagen and Delhi disagree, they work (their vocation requires it) to get to the bottom of the disagreement and determine whose view is better, more truthful. When Danish Lutherans and Indian Hindus disagree about theological matters, they typically stand their ground, largely ignore each

[9]According to press reports, Pope John Paul II is considering convening the leaders of the great world religions to discuss how they can contribute to the solution of pressing world problems, as he did in the 1980s. But this time around, unlike the last time, the Pope is considering not inviting His Holiness, the 14th Dalai Lama, because he is not the leader of a monotheistic religion.

other, and proceed as usual. Polkinghorne sees the religious reconciliation project as difficult. I see it as impossible unless some ecumenical council in the distant future waters down the essential claims of each of the world's great religions to the point that they lose their distinctive doctrinal content. We expect science, when in an epistemological situation of deep disagreement, to seek agreement and in the process to gain fine-grained knowledge of the phenomena in question. But religion is very different. A serious effort at achieving consistency and reconciliation among the world's religions would render most if not all of them largely unrecognizable as distinctive religions. The only plausible outcome I can imagine is something like the watered-down view of contemporary secular humanists. This, if true, is bad news for the sophisticated conciliationist.

There is a variety of sophisticated conciliationism that sectarian religious folk will find far too liberal. Contrary to Polkinghorne's insistence that the incarnate nature of Jesus Christ as God be preserved and that human nature be understood as lying between those of animals and angels, this sort of conciliationist will recommend that the diverse scriptures that are read as evidence for Yahweh, Allah, and a trinitarian God be read as referring at root to the One True God, Who is the same. Their distinctive moral content is to be understood as involving the attempt by diverse human groups to answer the question "How shall we live?" The differences among the conceptions of God and preferred moral conception are to be explained as due to contingent features of the people who produced the scriptures, and both are conceived as imperfect, adjustable, and modifiable by virtue of their imperfect origins. This sort of conciliationism is reasonable. But it will not appeal, at least not without much work, to the defenders and followers of sectarian orthodoxies—that is to most religious people.

Why am I talking about theological matters at all? Contemporary mind science discusses mind, agency, will, and the self, but it has nothing to say about God. Since my project is to explain how contemporary mind science suggests that we should understand these things, why should I get embroiled in the whole tendentious God issue? The answer is this: In the West, and, I think, in the East as well, the humanistic image within which views about human nature are held is inextricably tied up with views about the nature of God—his mind, his will, his agency, his nature (or, in polytheistic traditions, their natures). Different Gods yield different conceptions of human nature, and vice versa. Even the contemporary agnostic or atheist, who abides the dominant humanistic image of persons in what she insists is its secular variant, is tied by her heritage to certain threads that are embedded in the fabric that constitutes the theological history of the West, of Judeo-Christianity in particular. Thus we cannot search for common ground between the scientific and humanistic images without understanding the conflicts and tensions within the humanistic tradition itself.

Isn't It Ironic?

There is another strategy for avoiding conflict between the manifest and scientific images that deserves mention. This is to take an ironic attitude toward all ways of speaking, scientific and nonscientific. Minimally, being an ironist will require humility about the image you espouse since we know that images get displaced over time—Why should ours be any different? A more advanced ironist, employing even greater humility, might consciously acknowledge that the way she speaks and sees the world is a highly contingent matter, unlikely to capture the truth. She is amused by this and open to changing the image she operates with if need be, but she nonetheless proceeds to live as each person must, with some image in place, with some "final vocabulary," as Richard Rorty calls it, as her guide. A still stronger ironism tries to level the field among all forms of discourse by saying that they are all equally *subjective* or, what may come to the same thing, they are all equally *nonobjective.* Scientists talk about capturing reality as it really is, as do some philosophers and theologians, but the ironist sees this as just a form of boasting, a *façon de parler* with no real bite. No one is in a position to say how things really are. Of course, most people, scientists and nonscientists alike, will speak *as if* they know how things are, but if they do so without tongue in cheek, they need the sort of tonic only an ironist can provide.

I will say this much about ironism. Some of my best friends are ironists. I myself am an ironist when it comes to the manifest image to which I am attached—an attitude I recommend to everyone. I also recommend humility on the part of scientists. History shows that whatever scientists claim is true, at a later time is often proved false. But I think it is a big mistake not to believe that science tries, and sometimes succeeds, to understand how things really are. As for my ironist friends who think that science is no more objective than any other way of thinking, I have observed in most of them a fairly deep ignorance about science. Being around intellectuals who know almost nothing about science does not particularly bother me, except when they pronounce on the nature of science. My view is that if you are going to claim that all forms of discourse are equally subjective, you better have real familiarity with *all* the forms of discourse you aim to level. Otherwise, I strongly recommend following Ludwig Wittgenstein's advice: "Whereof one cannot speak, one must be silent."

Dueling Images

I began this chapter by talking about the importance of the human image as the ground from which we see our individual selves and within whose terms we find value, meaning, and a picture of our place in the universe. The image of what I called an in-group is the picture for that group within which a person's self-conception emerges and from the perspective of which the story of his life is to unfold—if, that is, it is to be a good life, as conceived in its terms. I argued that the image cannot be viewed simply as a conceptual scheme because it is not normally fully articulated in the minds of those who abide it. The image also functions as background, a set of largely implicit assumptions about what it means to be human. We are profoundly attached to the set of implicit and explicit beliefs that our reigning schemes may contain. It is this aspect that explains how an image can guide one's action, how it can give rise to a way of living one's life, and why challenges to it are disquieting.

Orthodox Christians typically don't marry orthodox Muslims. One reason is that they don't meet enough of them. But a deeper reason why marrying an orthodox Muslim might be a bad—or at least disquieting—idea for me (even though I don't practice any form of theistic Christianity) is that we would differ too much in how we envision a good life. It might be hard for us to agree about how to be a unit, a couple, as well as about how to raise children within the constraints of a consistent and stable image of who we are, and who we and they ought to be.

The distinction between the manifest and scientific images is, I argue, one instance of the more general problem of competing images. But it is the instance I am interested in. The manifest and scientific images make inconsistent claims that are endorsed as *true*, as revealing the way things *really* are.

So far I have laid out a framework for understanding how and why alternative images can clash and why these clashes can be so unsettling. I now turn to the ways in which the dominant manifest image of mind, agency, will, and the self are under assault by recent work in the human sciences. I believe the manifest image should yield to good science, if and when there is genuine conflict. So what is it that the human sciences tell us about mind, morals, and agency? Quite a lot, much of which is disquieting.

THREE

Mind

> Q: What is the soul?
> A: The soul is a living being without a body, having reason and free will.
>
> —ROMAN CATHOLIC CATECHISM

> The Astonishing Hypothesis is that "You," your joys and your sorrows, your memories and ambitions, your sense of personal identity and free will are in fact no more that the behavior of a vast assembly of nerve cells and their associated molecules.
>
> —FRANCIS CRICK,
> THE ASTONISHING HYPOTHESIS, *1994*

THERE IS A PICTURE OF THE HUMAN MIND that I'll call "the Cartesian picture," since René Descartes articulated it more clearly and defended it more ably than anyone before or since. The picture has two intimately related components.

The first is that the human mind does not occupy space—it is, as philosophers say, incorporeal or unextended. The mind receives information from the body about the body's own condition and about the external world, and it gives the body orders. But the two-way communication occurs between a physical substance and a nonphysical one.

The second component is that the mind's nature is such that we are all free agents. Descartes writes:

> But the will is so free in its nature, that it can never be constrained. . . . And the whole action of the soul consists in this, that solely because it desires something, it causes a little gland to

> which it is closely united to move in a way requisite to produce the effect which relates to this desire.

In Descartes's view, the mind performs psychokinesis with every voluntary action. It deploys its immaterial powers to move the material body. The mind itself, however, is self-moved. Suppose I experience lust. I experience lust because my body has reacted to another body in a certain way and it has passed this information via the pineal gland to my incorporeal mind. Unlike certain excessively moralistic types, Descartes thought that experiencing lust and other passions, anger and fear for example, was involuntary. Passions are passions precisely because we are passive in relation to them. But what we do about the experience is a matter of free agency. If we decide to act on the relevant passion we have decided one way, and if we resist acting on it we have decided a different way. Either way, the decision is free: We have performed a voluntary act and we are responsible for it.

St. Augustine was the ultimate party animal until his early thirties, at which point he changed his ways and became an exemplary moral person, a great philosopher, a bishop, and eventually a saint. We might say that Augustine was ruled by his passions until he saw the light in his early thirties. But according to the Cartesian picture, we would not mean that he *couldn't* control himself. We would mean that he chose not to control himself or chose to control himself badly. Likewise, when Paul Gauguin decided that his artistic project required him to go to Tahiti and thus abandon his family in France, he made a free choice. He could have done otherwise. No one and no thing, not even his commitment to his artistic project, forced Gauguin to make the choice he did. He chose Tahiti over family of his own free will. Whether that choice was smart or moral I leave to you.

The twentieth-century philosopher Roderick Chisholm puts the point about free agency, what he calls "agent causation," this way:

> [I]f we are responsible . . . then we have a prerogative which some would attribute only to God: each of us when we act, is a prime mover unmoved. In doing what we do, we cause certain things to happen, and nothing—or no one—causes us to cause those events to happen.

According to Chisholm and other libertarians—metaphysical libertarians, not political—free actions are "miracles" exempt from causal laws. A free agent has at her disposal all sorts of information about her desires and options, but when she makes a choice in light of that information, she does so with a will that is self-moved.

This picture of agency should be familiar. God, as Chisholm points out, is according to a certain standard conception the Prime Mover Himself Unmoved. God wills the world into being by a completely self-determined act of will. If God's will were moved to perform the ultimate creative act by prior causes—some Godly version of genes and upbringing—then creation was not the free act of an unmoved mover. Because we are created in God's image, we have similar powers of free agency, of a sort suitable for finite creatures. We can't choose to fly, unless we buy an airline ticket, but we can choose to resist lust, to become a doctor or philosopher, or to abide the principles of morality.

So the Cartesian picture has two components. First, each person has an incorporeal mind that interacts with her brain and the rest of the body. The interaction is bidirectional. In one direction, the body sends information to the mind about its internal state and that of the external world; in the other, the mind gives the brain orders to perform or resist performing certain actions.

Second, each mind is equipped with a faculty of free will. Neither what this faculty wills nor how it wills yields to scientific explanation. Science can explain what is physical and what obeys causal law.[1] But

[1]There is ambiguity in the concept of causal law, or better in the concept of causation, that I will speak about in the next chapter. The ambiguity has to do with the question of whether all causation is deterministic or whether some causation is indeterministic, where by indeterministic we mean that certain event transitions are genuinely governed by probabilities, that is, where there is no set of antecedents sufficient to cause one particular outcome. For now, in saying that something is obedient to casual law I mean that it is obedient to the principle of deterministic causation, roughly, the principle that everything that happens has a set of causes sufficient to produce it. To say that something is subject to causal law in this sense simply means that it is caused (in a deterministic way). In this view, there need not be any law of nature invoked to explain the event in question. Suppose I stand up and hit my head on an open cabinet door and get a lump on my head. I go to the doctor, who asks what happened, and tell him that I hit my head on an open cabinet corner. We both assume correctly that the cause of my lump was my head's

free will, as a capacity of mind, is not a physical phenomenon and thus does not abide the principle of causal determination. All physical phenomena are produced by prior physical causes that are sufficient to produce them, and when these sufficient causes occur the effect is necessary. It must occur as it does. But the will is not caused to will by any set of necessary and sufficient prior external causes. It moves itself. Or to put it differently: I deploy my will to choose as I choose. How exactly the will wields its magic is hard to state in a neat and comprehensible account—Chisholm's "miracle" is the best formulation we have. But the key is that when I choose to act in one way rather than another I make the choice, and no set of external or internal causes cause me to choose the act I choose.[2]

---

hitting the corner of the cabinet. If either of us was insanely curious we could explain why my lump is the size it is by gathering enough information about my body, its starting position, the speed at which I stood up, the position of the cabinet door, etc., and applying the law $f = ma$ to explain its size. Of course, we would never do this. We know that my hitting my head on the cabinet door caused my wound. And there are no laws in physiology about head lumps and cabinets. So all head lumps have causes that are sufficient to produce them. But these causes are multifarious. And there is no causal law that explains all of them. Cabinet corners can cause head lumps, as can car windshields, baseballs, coconuts falling from trees, and growths originating on the surface of the skull.

There are certain laws that apply with complete generality. The law of gravitation, the inverse square law, tells us that the "pull" of one body on another is inversely proportional to their distance from each other. If the masses of two moving bodies and the distance between them are known we can measure precisely the gravitational force between them. It is inversely proportional to the distance between them. The law of gravity is universal in the sense that it governs all bodies. Thus we speak of the gravitational force of the moon causing the tides and the gravitational forces among the planets and the sun causing the precise orbital pattern of each planet, and so on.

There are a few known laws in physiological psychology, for example, about the speed of nerve conductance. But there are no laws in psychology, no laws about the mind/brain, with remotely the generality of the law of gravitation. Perhaps there are no such laws. Even if this is so, it may still be true that each and every mental event has a (set of) cause(s) sufficient to produce it. Indeed, most mind scientists *assume* that this is the case. But more about that assumption later.

[2]The standard view is that I have something like complete control over what I *do*. An even stronger view would be that I also have complete control over what I *think* (am tempted to do or consider doing), as well as how I *feel* before and after I act. The stronger view is not widely held. That is, most people think they have something akin to complete control over how they act, but that many thoughts and feelings simply occur to you, cross your mind, and happen to you.

FIGURE 3.1

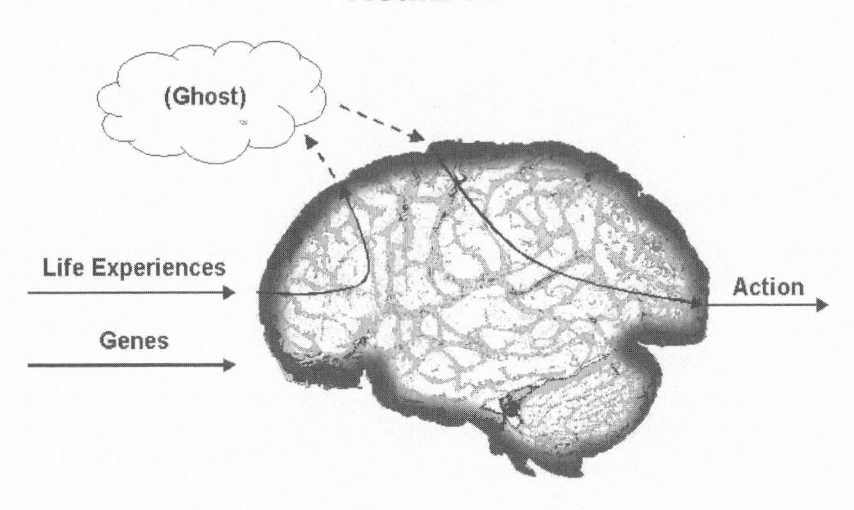

The Cartesian picture is still part of the dominant image of our age. Just fifty years ago, the philosopher Gilbert Ryle called it "the myth of the ghost in the machine," and described it as "the official doctrine." Most philosophers, psychologists, and religious leaders, as well as laypersons, subscribed to it. It still wields great influence, both explicit and implicit, over both lay and scholarly ideas about the mind. In our own time, world-class neuroscientists such as Wilder Penfield and Sir John Eccles have explicitly avowed a version of the Cartersian picture (Fig. 3.1).

This picture is in conflict with the picture of persons, mind, and agency projected by contemporary mind science. In fact contemporary mind science assumes the falsehood of both components of the Cartesian picture. Whether the relevant practitioners are prepared to explicitly reject the Cartesian picture, whether they even realize that their practice operates on the implicit assumption that the Cartesian picture is false, is a different, and in my experience, variable matter.

But if contemporary mind science is logically committed to the falsehood of the Cartesian picture, something has to give. Either the manifest image yields to the scientific image or the scientific image yields to the manifest image.

Peaceful Coexistence in the Marketplace of Images

There is a third possibility, which requires no one to yield, that I think best describes the current situation. This tactic is to avoid mentioning the whole mess, or to remain agnostic on whether it really is a mess.

This tactic, commonly used in dysfunctional families, invariably brings bad results. Dad is an abusive alcoholic. Mom is an enabler and in a state of constant fear and anxiety. The parents despise each other, and the kids are messed up. But the family makes believe things are okay. They only get worse.

Perhaps this analogy is not apt. Perhaps respectful silence will work in the end, or is already working. People think depression is not chosen. Whether the depression is brought on by some great loss or by the weather within, ordinary people now speak of serotonin problems in the brain as its proximate cause. Most everyone believes that taking a drink involves choice, but that if an individual has a certain genetic scaffolding she may become alcoholic. So alcoholism is conceived, at least in some measure, as a disease. To be more precise, alcoholism is conceived as an addiction, like cigarette smoking (and this because nicotine is addictive). Addictions are not diseases in the way HIV infection, malaria, or cancer are, but they are best explained in medical terms.

Explaining human behavior and mental states in terms of bodily states is widespread. Runners talk of dopamine highs. Horny male college students speak of testosterone surges, while their female classmates attribute much male behavior to testosterone poisoning. Guys meanwhile say of a female whom they perceive as "being bitchy"—acting like a female dog—that "she is having her period" or is suffering from PMS, or, more scientifically, that her estrogen and progesterone levels are out of whack. Lest you think that the war between the sexes is now being carried out at the level of sophisticated endocrinology, I hasten to point out that the endocrinological information being deployed is pretty primitive. Still, the coarseness of scientific understanding on which these ways of speaking are based is not my concern. My point is that these ways of speaking go deeper, at least in the minimal sense of saying more, than simply saying "he is an asshole" or "she is a bitch," and they do so by deploying scientific terminology.

There are many other cases where mind science has changed the way we speak and think about certain phenomena. Schizophrenia and autism are no longer thought to be caused by unloving, emotionally distant mothers, nor are many learning problems attributed to laziness or lack of motivation. Dyslexia, attention deficit disorder (ADD), and attention deficit/hyperactivity disorder (ADHD) have taken care of that. One in seven Americans is diagnosed with a learning disability, 80 percent of these with dyslexia. Dyslexia, according to our best current science, is thought to be due to disruption in the function of the brain areas—in particular in the angular gyrus—thought to subserve cross-modal processing, in particular, visual and language association. As for ADD and ADHD, perhaps they exist, but they are almost certainly overdiagnosed. If diagnosis continues at its present pace, then I recommend that at birth, when infants are given Apgar scores for health and vitality, all male babies simply also be identified as ADD or ADHD positive. This will save time and money. Gerber could then produce ritalin-laced strained peas for boys.

Kidding aside—kidding with a serious side—we can avoid cultural conflict between the manifest and scientific images by letting the scientific image gradually infiltrate and change the manifest image. This tactic, which appears to be already working, might be a relatively painless way for a manifest image to shift. Still it will find few fans among staunch and attentive defenders of the relevant manifest image, since the tactic presupposes that the scientific image will cause adjustments in the manifest image, not the other way around. The members of the Kansas Board of Education (many of whom were quickly voted out of office) saw this tactic for what it is, which is why they resisted teaching about the Big Bang and evolution in their schools.

Changing Words and Changing Images

One can sometimes detect a change in the manifest image by attending to the way people speak. As I noted above, we now talk of depression as a disease and alcoholism as an addiction, with the effect of taking some of the weight off weakness or misuse of "free will." Yet the mere fact that modern people engage in "neuro-speak" does not, by itself, tell us much about how they understand mind and agency. The majority of people now believe, if they have beliefs on the matter at all, that Copernicus's theory of planetary motion is true. But everyone still refers to the sun's rising and setting. Try asking someone on an early morning date to watch the earth's orbit bring the sun into visual range, and I will bet against the prospects for that romance. Does this mean that Copernicus's theory, despite being widely accepted as true, has not really sunk into our blood and bones? In a sense the answer is "yes."[3]

I suspect that most individuals do not see that there *is* a conflict between how they picture mind and agency and how contemporary mind science pictures these things, even if they engage in a fair amount of talk that indicates that they are absorbing a new way of thinking about mind and agency. Even psychologists, neurobiologists, and cognitive neuroscientists may not see how their projects are leading to the undermining and displacement of the "official doctrine."

Let me give an example. I co-taught a senior seminar at Duke University called "Mind and Brain" for several years with Gillian Einstein, a neuroscientist. The first year we taught the course I noticed that Jill sometimes went out of her way to say that some particular brain function, process, or outcome was *not* determined. I was puzzled because I thought most neuroscientists believed in some sort of causal determinism (I am not yet saying that they ought to). After many lunch talks, I came to understand that when Jill says some phenomenon is not determined, she simply means to deny *monocausation*. The APOE-4 gene, for instance, has a statistically significant relation to Alzheimer's disease. But people with APOE-2 and -3 genes also sometimes get Alzheimer's, and some old folks with the APOE-4 gene don't. Possession of the APOE-4 gene, despite being a common contributing factor, is neither necessary nor sufficient to produce Alzheimer's. But like most neuroscientists in my experience, Jill does think that any complex phenomenon has

[3]According to some polls, as many as 30 percent of Americans don't, in fact, know that the earth moves on its axis or in orbit around the sun.

some set of causes, however varied and complex, that are sufficient to produce it. Indeed, that they do produce it. Monocausation is rare, but mental phenomena are determined, even when we do not know every causal contributor.

Although it requires a bit of amateur psychoanalysis on a dear friend, my diagnosis of Jill's behavior is this. She knows that conceiving of the human mind deterministically is disquieting to students. Quite possibly it discomforts her personally. Despite believing that determinism is true, she says it is false. But in fact she only means monocausation is false. In this way the disturbing issue is glossed over, temporarily occluded from view.

Mind Science, Auto-Mechanics, and Anatomy

Everyone thinks that whatever an automobile does, or fails to do, has a natural cause, so too for the human body. Medical doctors believe that each and every medical function or malfunction has some set of causes that produce it. In both cases, the actual causes of a car problem or an illness can be notoriously hard to find. One reason is that each car and body has a unique history plus there are design differences depending on the auto manufacturer or on the genetic profile that is each human's birthright. Thus in auto-mechanic's courses and in medical school one learns "principles"—not what anyone would call "laws" in the strict sense that physics teaches laws. This is important. Although everyone thinks that cars and bodies obey the principles of causation—that for every event that happens there are causes operating at every junction—no one thinks it is a deficiency that we don't know, nor can we teach, strict laws of auto-mechanics or anatomy. One reason is that the unique history and design of the relevant systems mean that there are no absolutely general laws that apply to them—other than the laws of physics that everything obeys. The fact remains that we have very good knowledge in both cases of how the overall systems and their subsystems normally operate. And with this knowledge we are pretty good at tracking down the causes of breakdowns when these occur. Yet when an auto mechanic or a physician says that he just can't figure out what is causing some problem, he never says "perhaps a miracle occurred." Every auto mechanic and every physician assumes that causality is ubiquitous. Any unsolvable problem is a problem with his knowledge not with nature.

The quick way to put the lesson is as follows: Neuroscience is part of the science of anatomy. Its job is to describe how the brain functions in concert with the rest of the body and the environment. Seen in this way, the lack of strict laws in neuroscience is a consequence of the nature of the beast. The beast is us, *Homo sapiens,* and although there are many general system-level or functional "laws"—"the visual system works this way," "the auditory system works that way," "these systems interact with the motor system thus and so"—no sensible neuroscientist ought to tie the success of her science to the discovery of system-level laws similar in strictness or generality to laws like force = mass × acceleration. This law, in fact, applies to all physical systems including brains, but it is of no use whatsoever is explaining normal or abnormal function of such things as cars or human bodies.[4] Still, and this is the main point, every scientist applied or theoretical—just like every automotive engineer and mechanic—assumes that the system under investigation is obedient to the principles of cause and effect. One consequence is that the assumption of determinism—or, better, the assumption of the ubiquity of causation or of the principle of causal determination—does not entail that one believes or expects to discover strict causal laws of complete generality, nor does it entail that one ought to evaluate the success or maturity of one's science on the basis of making such discoveries.

So although Jill says something misleading if she is taken to be denying that each and every case of Alzheimer's disease lacks a set of causes sufficient to produce it, she is certainly right in denying that it is caused by one gene.

Causation and Volition

Actually the ubiquity of causation doesn't really bother people when it is things such as the workings of the visual system or the etiology

[4]Insurance adjusters and police investigators might use the "laws of mechanics" to determine the speed at the time of an accident. But no normal auto mechanic ever uses these laws, save intuitively when fixing a car.

of Alzheimer's disease that are being discussed. After all, the Cartesian picture acknowledges that the body works according to mechanical principles. The disturbing idea is that volition can be explained causally. The philosopher Thomas Nagel expresses the disturbing thought in these dramatic words:

> If one cannot be responsible for the consequences of one's acts due to factors beyond one's control, or for the antecedents of one's acts that are properties of temperament not subject to one's will, or for the circumstances that pose one's moral choices, then how can one be responsible even for the stripped-down acts of the will itself, if *they* are the products of antecedent circumstances outside of the will's control?. . . [T]he area of genuine agency, and therefore legitimate moral judgment, seems to shrink under this scrutiny to an extension-less point. Everything seems to result from the combined influence of factors, antecedent and posterior to action, that are not within the agent's control.

Many think this accurately describes how human agency must be conceived if we accept the scientific picture of things. I don't. What Nagel's statement does do—in an unwarranted and hyperbolic way—is call attention to the fact that the Cartesian picture of agency is in trouble if, in fact, causation is ubiquitous. And it expresses, again in a hyperbolic and anxiety-inducing way, a thought that disturbs many philosophers and ordinary people when they get to thinking about the problem of free will and determinism. But it is far from clear that situating human agency in the natural world requires that "the area of genuine agency . . . shrink[s] under this scrutiny to an extensionless point."

Consider this analogy: When I want to warm my house in winter, I place three or four logs in my woodstove and ignite them. If I make a good fire, it will burn for eight hours. The fire in the stove is (we'll suppose) a moderately long-term deterministic process. Every interesting fact about temperature and humidity change in my house during those eight hours will require mentioning, as the main cause, the fire in the woodstove. There is nothing in this mundane case of deterministic causation that requires thinking of the fire in the stove, or any other causal contributor, as shrinking to an "extensionless point."

The upshot is that even if some process or event is deterministic, its causal powers are in no way diminished. Even if we humans do not possess unconstrained Cartesian wills, it is not mandatory that we fall for the scary thought that agency is a will-o'-the-wisp. But many a good philosopher like Nagel has thought that the choice is between Cartesian free will and, well, nothing at all—that if there is no Cartesian free will, then human agency first shrinks to "an extensionless point" and then evaporates altogether. This is not the right way to describe the situation. Even if human agency cannot be explained in terms of a miraculous, otherworldly faculty of unconstrained will, we humans might—just like my woodstove—by virtue of our complex nature, abilities, and staying power, be major contributors to both our own lives and that of the world around us. So even if the Cartesian picture of agency goes, the evaporation of agency itself is hardly inevitable.

The Cartesian Picture Is Prescientific

Someone might accuse me of overstating the conflict between the Cartesian picture of mind and agency and the scientific image, and thus of trying to cause trouble where none exists. The claim would be that the Cartesian picture is a sophisticated articulation of the manifest image circa the mid-1650s, but it captures a prescientific picture that modern persons, having already absorbed a heavy dose of how science sees mind and agency, no longer feel in their bones. This response will not work. Ryle was right that in 1950—three hundred years after Descartes made explicit and defended a richly textured account of the image that had served as background for several thousand years—the image he articulated was "the official doctrine." And with *USA Today*, CNN, *The International Herald Tribune*, and *The New York Times* as my witnesses, most people today—at the beginning of a new millennium—believe strongly that the Cartesian picture is true.

Furthermore, Descartes's writings are a difficult body of work to label as prescientific. Descartes was a world-class mathematician as well as a world-class philosopher. What he didn't know he invented, analytic geometry for example. And he strove in many areas to apply precisely the kind of rational, mechanistic theory-making that has made science so successful. If Descartes's ideas are "prescientific," when, exactly, did science begin?

When Mind Science Began

This turns out to be an interesting question. If by *science* we mean the activity of distinguishing between appearance and reality, then it began very early—when the thought first occurred that the moon, despite appearing small, might not in fact be small, or when Democritus first proposed that solids are constituted of indivisible and imperceptible particles. If we mean by *science* the experimental activity of trying to confirm or disconfirm a hypothesis, then it began later. When Copernicus proposed his heliocentric theory of planetary motion in 1543 (it had, by the way, already been proposed in the fifth century B.C.E. by Greek astronomers), it was primarily because he saw that it was mathematically simpler than Ptolemy's geocentric theory. But when Galileo peered through a telescope in 1610 he provided experimental support for Copernicus's view.

The subordinate question of when the human sciences began is also interesting. Using the taxonomy I just proposed, we might say that the human sciences began when humans first distinguished how people seem from how they really are. This sort of inquiry began very early. All the greatest religious and philosophical texts—the Koran, the Puranas and the Bhagavad Gita, the Upanishads, the sacred writings of Tibetan Buddhism, the Old and New Testaments, Plato's *Dialogues* and Confucius's *Analects,* to name just a few—take human nature as one of their central topics. One might agree that inquiry into human nature began in ancient times, but still wonder when experimental human science began.

IN COLLEGE, I WAS TAUGHT that experimental psychology was invented by Wilhelm Wundt in Leipzig in 1879. This now strikes me as an odd claim. To be sure, Wundt brought subjects into rooms and used reaction time devices. But his primary experimental method was introspection—asking subjects to respond to some stimulus when they saw it or heard it in a certain way. And granted, Wundt's introspectors were "trained" introspectors—but this largely meant that he asked people who gave "incorrect answers" to leave his lab. But introspection had been a favored technique in the inquiry into human nature from the beginning of time.

Putting aside for a moment the slew of objections to introspection as a reliable technique to get at the nature of mind, there is another way of defending the idea that scientific psychology was invented in Leipzig one hundred twenty-odd years ago. One may say that it was only in the late nineteenth century that mind science saw its role as simply describing the mind and human nature, and began trying to confine itself to description and explanation.

When Socrates tells us that it is good to "know thyself"; or when Aristotle says that "true friendship"—friendship based on genuine and abiding love and respect for the other—is better that friendship based on using others solely for one's own pleasure; or when Confucius extols the virtues of filial respect and piety; or when Buddha teaches that certain emotions are destructive and worth the effort of overcoming, of transcendence; or when Jesus says that we "ought to love our neighbor as ourself," they are doing more than describing. They are recommending a certain way of being, of manifesting one's identity. And recommending, it might be claimed, is not part of science.

Suppose we provisionally accept this last premise. Still it is not obvious that the recommendations for living made by these great thinkers are not in some sense experimental. Jesus, Buddha, Confucius, Socrates, and Aristotle make their recommendations based to some (possibly large) degree on empirical observation. They are, insofar as they are engaged in cognitively respectable ethics, doing what I earlier called ethics as human ecology. Lives lived according to the ideals they propose go better, indeed are better, than lives lived according to the statistical norms. How could they know this? Observe people who live both ways and see who flourishes. Water your plants. They will do better than if you don't. Eat breakfast. You will feel and think better than if you don't.

The ancient Egyptians built aqueducts of stone with rectangular arches. By 1400 B.C.E. they realized that this was not the best design.

The evidence was provided by the fact that aqueducts built this way collapsed.

The reason is that a slab mounted on pillars has all its weight distributed downward. Romans took this evidence to heart and built aqueducts with rounded arches, which distributes weight outward into the banks of the river. Stone is much better at resisting crushing than shearing forces. Thanks to the Roman architects, I was able in 1990 to walk across the Pont du Gard, an aqueduct and viaduct built in the first century C.E. that still stands in the south of France.

I see no interesting difference between the recommendations of the great ethicists for how best to live and those of architects for how best to build bridges or aqueducts. Both are based on empirical observation relative to certain considered ends or goals.

I am not suggesting that ethics or architecture can deductively derive oughts from is's.[5] That would be logically impossible as David Hume famously taught us. What you can do, however, is specify and defend certain goals and ends and then show inductively the way of being or building that will best meet those ends.

Just because psychology prior to the last one hundred and twenty years was not experimental in our sense—there were no government grants, paid subjects, or brass instruments, no computers, no laboratories inside buildings, no double-blind experiments—it does not follow that this earlier psychology was not empirical, that is, that it was not based on making surmises from experience. Nor does the fact that this psychology was prescriptive through and through distinguish it from applied mind science in our time. Psychiatry, psychotherapy, and neurology assume certain norms of healthy functioning and prescribe methods—sometimes pharmacological,

[5]You can deductively derive the conclusion that an aqueduct built according to Egyptian design will collapse faster than one built by Roman design. But only if you import the premise that it is good for bridges to last will you gain the conclusion that you ought to use Roman principles. And the argument for that premise will come from the specification of certain goods and goals that cannot themselves be derived from any set of what we call "objective facts." But no sensible person will complain about importing the required normative premise in this case. Why is that? Because the goal of aqueduct building is uncontroversial. It suits a clearly recognized end, a sensible goal. Thus it is something that *ought* to be built.

sometimes not—to achieve or maintain them. If you are depressed, perhaps you ought to take Prozac. If you have a pituitary tumor, you ought to take bromocriptine or have surgery unless, that is, you don't care about being a hormonal disaster site or losing your eyesight. If family relations are poor, counseling might help.

The Order of the Sciences

If I am right about this, then something is wrong when Bertrand Russell writes:

> The sciences have developed in an order that is the reverse of what might have been expected. What was most remote from ourselves was first brought under the domain of law, and then, gradually what was nearer: first, the heavens, next the earth, then animal and vegetable life, then the human body, and last of all (as yet very imperfectly) the human mind.

If we grant that the great early ethicists were engaged in empirical mind science—both descriptive and explanatory inquiry as well as human ecology aimed at uncovering how we ought to live in order to flourish—then empirical mind science was one of the first sciences. But if Russell means that the experimental techniques taught at universities in "Methods" courses were followed imperfectly by early mind scientists, then he is right. And he is most certainly right when he speaks of bringing the objects of inquiry "under the domain of law." Mind science is at a very early stage of doing that. It remains possible—indeed I would say likely—that even if everything is caused, we will never bring the most interesting causal sequences under strict universal causal laws.

Over a century ago in 1890, William James published a two-volume compendium of every known discovery in mind science—*The Principles of Psychology*. Two years later, in a shorter version of *Principles* aptly entitled *Psychology: The Briefer Course,* here is how James described the state of the art: "A string of raw facts; a little gossip and wrangle about opinions, a little classification and generalization on the mere descriptive level . . . but not a single law in the sense that physics shows us laws."

Are things different now? Yes. There are many promising theories in mind science and a few mostly local laws, nothing remotely as general as force = mass × acceleration or $E = mc^2$. Invariably these are statistical laws rather than strict deterministic ones. For example, it is probable to some value *P* that a person with the APOE-4 gene will develop Alzheimer's disease before age 70, where *P* has a significantly higher value than it does for a person with the APOE-2 or -3 gene. There is no interesting law in mind science where the probability of some outcome is assigned a value of 1 (that is, certainty). This is likely due to the extraordinary complexity of the phenomena being studied. It is not that neuroscientists think that the mind/brain disobeys the principles of cause and effect.[6]

Contemporary mind science holds great promise for revealing the deep structure of mind. But it would be dishonest to give the impression that it has already delivered the goods. No systematic and unified theory emerged in the "Decade of the Brain," a designation for the 1990s effectively lobbied for by The Society for Neuroscience, a society that has seen its membership soar from 500 in 1970 to over 25,000 in 2000.[7] That said, neuroscience is, by all lights, an extreme-

[6]The fact that a law is statistical does not imply that an outcome is not strictly determined. Consider the law that says that a toss of a fair coin has a 50–50 chance of landing heads or tails. Which one occurs is believed to be completely determined on each toss by the set of causes consisting of its original position, the speed generated by the flipping finger, etc. Of course, we are almost never in a position in advance, or even if we try to track the relevant variables as the toss is made, to specify all the relevant variables. One important lesson is that even if some event is determined and even if we know the determining laws—in this case these include f = ma and the law of universal gravitation—we may not be able to predict the outcome precisely. We might be able to explain why one outcome occurred after the fact, but the important point is that explanation does not entail prediction. Remember this. It is important.

[7]I spoke on "Free Will and Neuroscience" at a plenary session of the Society of Neuroscience at the Anaheim Convention Center in 1993. Over a thousand people were in the audience. While speaking, the thought passed through my mind that the way I was feeling must be something like Mick Jagger feels when he performs, which led me to wonder, with some pleasure and trepidation, how I would deal with the groupies when I was finished. My speech was reported by the Associated Press, picked up by various newspapers, even leading to a full-blown editorial in a paper in Des Moines. Things like this don't happen when I speak at American Philosophical Association meetings. It is pretty clear where the action is perceived to be.

ly promising and progressive research program, one that can hopefully be expanded on and drawn into some sort of systematic and unified picture. Degenerative research programs go nowhere, yield no testable predictions, and yield no knowledge. Astrology is an example. Progressive research programs give rise to testable hypotheses, some of which are confirmed, and lead to further testable hypotheses. New knowledge is discovered. Ignorance is diminished.

Regulative and Constitutive Ideas

Immanuel Kant drew a distinction between the ideas that regulate or guide an inquiry or project, and those that are the result of an inquiry or project so guided. The first kind he called regulative ideas (or ideals), and the second he called constitutive ideas. Most people who get married assume that the marriage will last and regulate their behavior so that it does. The assumption is thus a regulative idea. If the marriage actually lasts, this is a constitutive fact about it. It really happened—the marriage worked.

Most scientists assume that the world is obedient to "laws," which they sometimes discover. But the regulative assumption that such laws exist dramatically outstrips the number of occasions on which they are actually discovered. However, when laws such as Newton's laws of motion are discovered, they become a constitutive part of science.

Even if a field of science is not doing very well in discovering laws, it may be discovering facts and regularities. Imagine a time when nothing was known about the brain besides its appearance and location. Scientists set to work assuming that they could learn some facts about this *terra incognita*. This simple assumption—that there were facts waiting to be discovered—regulated their inquiry. But they possessed no knowledge about the mind/brain. The book of constitutive facts was blank. Time passed and the pages began to be filled up. The brain looks a lumpen grayish mass but actually contains on the order of 100 billion neurons, that is, at least 10^{11} neurons, possibly as many as 10^{14}. These neurons communicate with other neurons at junctures called "synapses" (by way of lines of input and output known as dendrites and axons). Each neuron has synaptic connection with, on average, 3,000 other neurons, producing approximately 100 trillion synaptic connections. In addition, the brain contains different structures with different functions—Broca and Wernicke's areas for language, the limbic system for emotional processing, frontal cortex for planning. It is also awash with chemicals such as acetylcholine, serotonin, dopamine, and norepinephrine. This whole system is connected with great intricacy to the spinal system and to the eyes, ears, skin, fingers, and toes.

That one makes a regulative assumption does not mean it will pan out. The brain could have turned out to be remarkably simple, so that the assumption that there were interesting facts to be discovered about it could have proven false. At the end of their inquiry neuroscientists might have been pretty much where they started: between the ears is a homogeneous piece of meat weighing on average 2 1/2 pounds. Its function? It cushions the skull.

The distinction between regulative assumptions and constitutive ideas, facts, outcomes, and laws is useful in thinking about marriage, investment in the stock market, planning a trek, science, and even literature. My old friend Stanley Fish gives a nice example of the distinction in a humorous op-ed piece about the sad state of contemporary biography that appeared in the September 7, 1999, issue of the *New York Times*. Fish writes:

> [O]nce upon a time biographers didn't have to invent connections because they came ready made in the form of master narrative models. The two most durable were the *providential model*—everyone lives out the pattern of mistakes bequeathed to us by the original sin of Adam and Eve; and the *wheel of fortune model*—every life worth chronicling is an example of the principle of the general rule that what goes up must come down.

One might say, speaking as a Kantian, that the *providential model* and the *wheel of fortune model* regulated the behavior of biographers. The subject's life is seen as lived according to the terms set by the providential or wheel of fortune model, and the biography written accordingly. Fish contrasts this with the contemporary "undisciplined" biography, which simply depicts a life as a string of unconnected contingent facts. But perhaps these books are not undisciplined but instead are following a new and different regulative model that conceives human lives as in fact consisting of strings of unruly facts and events.

One consequence of writing biography from the perspective of a particular master narrative model is that the life being told will conform to the model. Will the life really have been lived that way? An advocate of a particular regulative, master narrative model might not be much interested in discussing this question—the question of whether the way he frames the life he discusses is the right way. It is. It must be. Why would he have committed his life to studying things from this perspective, rather than some alternative, if he did not think it had the potential to capture things as they really are?

One might worry that the regulative ideals that set the terms for an inquiry—be it biographical, scientific, or philosophical—are not themselves testable. This is a legitimate worry. But it is partially mitigated by the fact that a regulative ideal can lead inquiry in productive or unproductive directions, thus providing feedback about the ideal's wisdom, or at least its applicability to a particular subject. Vitalism in biology, which posits immaterial spirits, proved less productive than its materialistic alternative. And despite their rarity, there do seem to be some life stories that don't abide the strictures of either the providential or wheel of fortune models, as even advocates of these models must admit.

There is one glitch in the idea that the regulative ideals guiding an inquiry are indirectly testable by way of feedback about their success. The success of a research program will be determined by what the relevant theorists have decided in advance constitutes success. From the point of view of an advocate of the Cartesian picture, any diminishment of the space of free will indicated by advances in mind science might be seen as revealing that the research program is unprogressive, for it is headed in the direction of denying what we know in advance is true, namely that we are free agents. There is a way around such conflict, and it involves foundational inquiry, examining the coherence and logic of the underlying assumptions themselves. Philosophers are in this sort of business. And I hope not to be leaking sad or bad news in telling you that the vast majority of contemporary philosophers think that the Cartesian picture describes the mind in an unstable, possibly incoherent way.

The Regulative Ideas of Contemporary Mind Science

What are the regulative ideas that guide contemporary mind science? I have indicated that there are at least two that involve denying the two components of the Cartesian picture. The first regards the nature of mind; the second the nature of agency. The best compendium of the state of contemporary neurobiology, by Dale Purves and his colleagues, prefaces itself this way:

> Whether judged in molecular, cellular, systemic, or behavioral terms, the human nervous system is a stupendous piece of biological machinery. Given its accomplishments—*all the artifacts of human culture, for instance*—there is good reason for wanting to understand how *the brain* and the rest of the nervous system work. (italics added)

Contrast Purves's quote with this alternative:

> Whether judged in terms of its nonphysical nature, or in terms of what it accomplishes—it is the source of meaning, free action, morality, and human culture—*the mind* is a stupendous thing. There is good reason for wanting to understand how the mind works.[8]

Is there any meaningful difference in these two ways of posing the question? Yes. Descartes could have easily prefaced his *Meditations* in the second way, but not in Purves's way. In fact they are not really asking the same question. Modern mind science regulates its inquiry by the assumption that the mind is the brain in the sense that perceiving, thinking, deliberating, choosing, and feeling are brain processes. The manifest image, insofar as it posits that we can understand mind bet-

[8]The word "understand" will need to be read differently in the Purves quote and the Cartesian analogue. Understanding for a neurobiologist means to understand as a natural phenomenon. Since the Cartesian does not think the mind can be understood that way, he will mean something like "render philosophically intelligible."

ter than we now do, thinks the brain is at most a conduit for a mind that is much more than just a piece of biological machinery. In its view, the mind is not part of the world to be revealed by physical science because it does not entirely belong to the physical fabric of things.[9]

That the mind is the brain is thus a regulative assumption that guides contemporary mind science. There is a second assumption as well.

In his 1917 book *Introductory Lectures to Psychoanalysis,* Sigmund Freud wrote that mind science assumes determinism, that each mental act, be it conscious or unconscious, has psychobiological and environmental causes that are sufficient to produce that act, indeed make it inevitable. When we privilege conscious causation or will we underestimate the role of other causes, including the unconscious causes of the conscious mental states themselves—the most significant underlying causes of the state of our wills. Freud acknowledges that the thesis of determinism, the assumption of a causally closed universe, "offends against an intellectual prejudice" as well as "against an aesthetic and moral prejudice." He immediately adds: "We must not be too contemptuous of these prejudices; they are powerful things, precipitates of human development that were useful and indeed essential. They are kept in existence by emotional forces and the struggle against them is hard."

But however hard the struggle against the opposing prejudice, the deterministic assumption is ubiquitous in mind science, throughout good research programs and bad. Yet an assumption may be conceptually ubiquitous, even conceptually necessary, even though those who hold it are not fully aware that they do so.

Another possibility is that most mind scientists assume determinism and are aware of this, but disclaim the assumption because they sense that it will not be well received by people they wish to gain as allies. The "Grand Inquisitor" chapter of Dostoyevsky's *Brothers Karamazov* is disturbing precisely because it raises the possibility that the

[9]Purves's book contains no index or glossary entry for mind, free will, or determinism. Indeed, the only entry with the word "mental" in it refers the reader to the pages in which mental retardation is discussed as a psychobiological disorder.

defenders of the faith—the cardinal and his clergy—may in fact be atheists, despite avowing, for public consumption, the opposite. Yet another possibility is that mind scientists do assume something like the principle of universal causation when engaged in mind science, but don't see how this assumption has implications for their general allegiance—outside the lab, as it were—to the traditional notion of free will.

Circumspection and Free Will

We may confidently dismiss the bad faith scenario. But the first and third scenarios, in which mind scientists don't quite see what their behavior or their logic commits them to, might be true. The word "circumspection" has its roots in two Latin words meaning, when compounded, "to see around." (One of the great blessings of Catholic high school education was four years of Latin and two of Greek.) In English, the word "circumspection" is normally used in contexts where discretion is called for. You should be circumspect when you don't want to call attention to what you are doing—when you want others to see around you or past you. One might use the term (although no one does) to refer to first-person "seeing around." Sometimes when I can't find my keys they are right in front of my eyes. But I look past them, see around them. It is commonplace to say that people just don't see their own faults or shortcomings—even when these are, as it were, right before their eyes.

It is possible that both kinds of circumspection occur when the problem of free will surfaces.

In the summer of 1978, I gave a talk at the American Psychological Association Meetings in Toronto. The topic was the implications of sociobiology (now called "evolutionary psychology") for the sciences of the mind. I accused E. O. Wilson, the founder of sociobiology, of stretching things in claiming that sociobiology would absorb psychology, but I suggested he was most certainly right that mind science would benefit from fully taking in Darwin's lesson that we are animals with no magical powers—such as Cartesian free will.

It was my first professional talk and I was very nervous. I also remember being very surprised that several of what we would now call cognitive neuroscientists in the audience were disturbed that I questioned the existence of free will. Being young and naive, I was extremely puzzled at this. I asked: "Don't you assume that each and every human motive, thought, or action that you are interested in explaining has some set of causes sufficient to produce it?" This question produced the most extraordinary mental gymnastics. As far as I could discern, the answer to my question was "Yes, I (we) do assume determinism when we study some psychological phenomena, but don't make too much of that. Humans obviously have free will."

In the past twenty years I have gotten used to this response, having heard it again and again from my undergraduate and graduate students in psychology and neurobiology. Philosophy graduate students have little trouble in seeing the problem with this type of circumspection—seeing around or past what you are logically committed to, denying the validity of one of your regulative ideals—since discussing this sort of thing is everyday fare in our seminars. But it remains bewildering to me that, when pressed, budding mind scientists will acknowledge that they assume determinism in their practice as mind scientists, but that they also don't assume it could really be true. Shades of William James.

Some mind scientists clearly see that this position is unstable. But I am not sure they see clearly why if the traditional conception of free will is not a credible assumption in the lab, it is not a credible assumption outside the lab either.

Let me discuss a specific example. Antonio Damasio, one of the most philosophically sophisticated and eminent cognitive neuroscientists of our time, asks of his own neurophilosophical theory of the person: "Does it deny free will?" He answers "no"—humans have free will. But it is clear that Damasio doesn't believe that we possess anything like Cartesian free will. The "free will" Damasio has in mind is a conception involving what he calls "suprainstinctual decision making capacities" that are acquired through the interaction of a conscious embodied being with the natural and social environment and that are to be explained in fully naturalistic terms.

In his best-selling book *Descartes' Error,* Damasio explains the book's title as follows: "This is Descartes' error: the abyssal separation between mind and body." And he claims, just as I have, that "Descartes' error remains influential. For many, Descartes' views are regarded as self-evident and in no need of reexamination." Damasio sees this error as one error, whereas I see two mistakes: (1) the claim that our minds are disembodied and (2) the claim that our wills are free.

It might be fairer to claim that Damasio thinks that "the abyssal separation between mind and body" yields free will as a direct consequence—and thus if one is overturned, the other falls too. But it is not true.

Here is a true story that reveals why it is wisest to think of Descartes's philosophy as containing two distinct theses regarding the mind–body relation. Two major seventeenth-century philosophers, Gottfried Leibniz and Nicolas de Malebranche, tried mightily to solve the greatest difficulty facing Descartes's theory of free will, namely, the problem of mental causation. How can a nonphysical mind cause a body to move? How can information about the external world be received by our senses and cross the line that divides the physical and the mental so that we consciously perceive things outside us—sunsets, trees, birds, toads, butterflies, flowers, and stars? Both Leibniz and Malebranche settled on a view that involved preserving mind–body dualism while abandoning Descartes's account of mental causation, which involved bidirectional causal interaction between the mental and the physical. Since the Cartesian conception of free will, the one that is to this day part of the manifest image, requires such interaction, Leibniz and Malebranche are best understood as accepting Cartesian dualism but rejecting the Cartesian conception of free will.

If Descartes's error is, as Damasio claims, all of one piece, then it should not be possible to accept dualism while rejecting interaction between mind and body. But both Leibniz and Malebranche saw correctly that the thesis of dualism, the claim that persons partake of two realms—one mental, the other physical—was logically independent of the thesis that these two realms interact. Since the Cartesian conception of free will turns entirely on mental events, choices, causing bodily events, the rejection of the idea that the mental and physical interact was tantamount to rejecting the Cartesian account of free will. How is it possible to believe in Cartesian dualism while rejecting the Cartesian conception of agency? Here's how.

According to Leibniz, God, in creating the universe, set out a plan for each person's mind and body to run in parallel. When I consciously decide to go to the ballgame and go, it is not that my decision causes my body to get up and go to the game. Rather, God preestablished a harmony between my mind and my body. It seems as if my decisions move me, but this is not so. At creation, God established a master plan of unbelievable complexity that included my life. He made things so that my nonphysical mental nature and my physical nature run in parallel. Save for God's initial creative act *ex nihilo*, the problem of mental causation is solved by keeping the mental and the physical from interacting (and the initial creative act of making Matter from Mind or Spirit *ex nihilo* is not a problem since it is God who did it). Mental events can give rise to other mental events. I *decide* to go to the game and then *enjoy* myself. And bodily events can give rise to other bodily events. I *go* to the game, *eat* popcorn, start *digesting* it, and so on. But mental events and bodily events do not interact.

Malebranche's view differs from that of Leibniz mainly in that it makes God busier, which, for God, is not a serious problem. Where-

FIGURE 3.2 Two Ways God Coordinates Mental and Physical Events

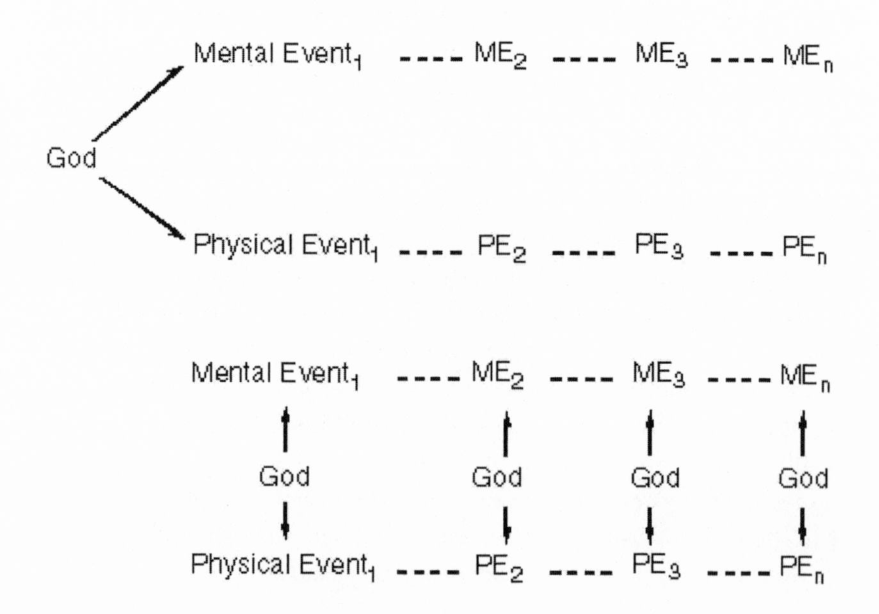

as, in Leibniz's view, the mental and bodily clocks are set in harmony at creation, Malebranche has God make sure they stay aligned by intervening on each and every occasion (Fig. 3.2). At the exact moment I decide to go to the game God gives my body "the get up and go" order. When I am at the game and find myself enjoying it, God produces certain changes in my bodily "humors." Once again the mind and body, despite seeming to interact, don't. God sits in the middle and produces mental changes in one direction and bodily changes in the other direction. Of course, the stadium is full of people, all engaged in harmoniously coordinated thought, feeling, and action. But it is God doing the coordination, not each individual.

Parallelism is the view that at the moment of cosmic or personal creation, God preestablishes harmony or synchrony between the mind and body of each person. *Occasionalism* is the view that on each and every occasion where there seems to be causal interaction between mind and body, there is no such interaction. It is God who causes the mind and the body to behave harmoniously. In neither view do mental events and bodily events causally interact. Indeed, neither view requires causation even within the mental and bodily realms. God might cause the harmonious relation between mind and body with the right mental and bodily events simply temporarily following each other in the right sequence. However, parallelism, in particular, allows three kinds of causation: divine causation creating mind and body and preestablishing their harmony; mental causation of the sort that operates when a mind deliberates and produces a decision; and physical causation of the sort that causes the pupils to contract to light or that produces the series of physical events that constitute driving a car to some destination. Whereas both views insist on the immateriality of mind, they are compatible with determinism—with the denial that humans have free will.

I draw attention to the logical independence of the thesis of mind–body dualism from the thesis that there is bidirectional interaction between the mental and the physical not because I think either Leibniz's or Malebranche's view is plausible. They are not. Indeed, the fact that Descartes's view that my nonphysical will causes physical things to happen remains dominant shows that these other views could not, for a variety of good reasons, grab the attention of the defenders of the manifest image. I mark the logical independence in order to point out that Descartes's error, if he is guilty of one, need not be diagnosed in the way Damasio does. Leibniz and Malebranche, unlike Damasio, saw dualism as unproblematic—indeed, they made the abyss between mind and body even more "abyssal" than Descartes did—but saw Descartes's account of mental causation as erroneous.

A third possibility is that both theses are erroneous. As I read Damasio, he thinks Descartes's one error—the abyssal separation—is logically of one piece, so that rejecting dualism logically entails rejecting Descartes's conception of free will (while still insisting that free will exists in some non-Cartesian sense). But this assumption of strict entailment is wrong, as the theories of Leibniz and Malebranche reveal. Immaterialism about mind does not entail that such minds exercise free will. In fact, on some interpretations of parallelism and occasionalism, immaterialism sits comfortably with causal determinism, even if it is God who does all the causing.

Damasio could respond that rejecting dualism, insofar as it involves conceiving of mind as a fully embodied part of the natural world, does in fact entail the rejection of Cartesian free will. Maybe, maybe not. This is still an issue of contention among some philosophers and mind scientists. Some think that if you naturalize mind, then given that natural things obey strict causal laws, there is no room for Cartesian free will. Others can point out that the phrase "*given* that natural things obey strict causal laws" makes an additional assumption beyond the claim that the mind is embodied, part of the natural fabric of things. This assumption is not self-evident, and the behavior of particles at the quantum level seems to show that it is not universally true. *Some* things obey statistical or probabilistic laws.

Thus even if we reject Cartesian dualism and accept a naturalistic picture of mind, free will remains an open question until we fully understand the nature and types of causation that exist, and of those, the types that govern mind. The mind/brain might work in some yet-to-be-understood way that is not deterministic and that preserves something like the Cartesian idea of free will.

Several scientists, most notably the British Nobel Laureate Roger Penrose and Stuart Hameroff of the University of Arizona, speculate that quantum physical properties of microtubules in neurons hold the key to explaining consciousness. And many mind scientists think that some combination of chaos and complexity theory and the theory of self-organizing dynamical systems is needed to explain mental causation. These theories, or some versions of them at any rate, claim to work best at explaining the behavior of phenomena that do not obey strict deterministic laws.

Whether such theories have resources to save something like the Cartesian conception of free will remains to be seen. For now the key point is this: Descartes's two theses are logically independent. Within our everyday humanistic image, they go together like horse and carriage, but—contrary to the old tune—you can have one without the other.[10]

I emphasize this point because it calls attention to the strategic question of how best to go about changing the manifest image. In the spring of 2000, I taught a mixed undergraduate and graduate level course on the philosophy of mind as a visiting professor at Boston University. Among other books, the students read Damasio's *The Feeling of What Happens* (the sequel to *Descartes' Error,* which many had also read) and Daniel Dennett's 1984 classic, *Elbow Room: The Varieties of Free Will Worth Wanting.* They enjoyed both books greatly, but Dennett's made them nervous in a way Damasio's didn't. Why? Because, although both argue for a fully naturalistic conception of mind, Dennett explicitly asserts that Cartesian free will is a variety of free will that (a) is not worth wanting and (b) you can't have even if you do want it because it is based on philosophically and scientifically incredible ideas, and is therefore incoherent, impossible. Over several weeks of the seminar I heard the refrain—"But, but, Dennett's conception of free will is not *really* free will!"

It is specifically the Cartesian conception of free will that is part of the manifest image and thus the only contender for an acceptable view of free will. It is *the* variety of free will most everyone wants and thinks they need. So despite Dennett's compelling arguments that we should

[10]For the youngsters reading this book, there was a song I listened to when I was a boy that had these lyrics: "Love and marriage, love and marriage. They go together like a horse and carriage. You can't have one, you can't have one, you can't have one without the other." Good lyrics, but logically, even empirically, false.

not want this incoherent variety of free will and that there is, happily, a type of naturalized free will we can and do have that gives us pretty much everything we need, he gave my students the heebie-jeebies.

Neither of Damasio's books got this response—not because he doesn't hold the same sort of view as Dennett but because he is circumspect about it. When Damasio says free will is real, he definitely does *not* mean Cartesian free will. But because the Cartesian conception of free will is the default view within the manifest image, he is not read as challenging that view. The words "free will is real" are read as comforting, as meaning that Cartesian free will is real, even though when coming from Damasio's pen, they definitely do not mean that.

Admittedly, there is a strategic issue here. Earlier I spoke of a range of alternative ways to change the manifest image of an age. There are basically three tactics: (1) You may say, "You are a fool to believe ϕ, we'll have to straighten you out." (2) You may say, "ϕ is not true, but ψ is and ψ is close enough to ϕ that you'll eventually get used to it. Trust me, everything will be alright." (3) You speak of ϕ meaning ϕ, while I speak of ϕ meaning ψ hoping that my way of speaking will eventually win the day in this play of language games, so that you, or your descendents, will eventually come to mean ψ when you or they say ϕ.

Dennett deploys sometimes the first and sometimes the second tactic, while Damasio uses something like the third tactic when discussing free will but the first and the second on dualism. Which tactic is most effective is not entirely clear. But that is a matter for intellectual historians and sociologists of science to settle.

When Regulative Assumptions Become Constitutive Ideas

The fact that even among mind scientists there is a certain amount of circumspection with respect to the assumption of determinism (versus free will) is evidence that the idea that the mind is the brain has moved more firmly from regulative to constitutive than the assumption of causal determinism has. Wittgenstein noticed that this sometimes happens: "The same proposition may get treated at one time as something to test by experience, at another as a rule of testing."

To make the point ever so briefly, the assumption that the mind is the brain is panning out experimentally. Some will say that all that mind science has produced are discoveries of correlations between mental events and brain events. One favorite philosophical thought experiment to show this goes like this: Imagine I hook you up to a cerebroscope that allows you to watch your brain-in-action on a TV screen. I show you paint chips of many different colors and we both watch as your brain lights up when you see red, yellow, green, blue, turquoise, magenta, and so on. You say "Lo, red!", "Lo, blue!" and so on. For each experience you and I both see an image of your brain activity on the cerebroscope. I say: "There you have it. You've now seen your own experiences of different colors." To which you stubbornly reply: "No, I haven't. At most I've seen images of what my brain was doing when I experienced different colors."

You are right, of course, in one crucial sense. The images revealed by the cerebroscope are not the experiences themselves. The images are on a TV screen, whereas the experiences are in you. I claim, however, that the activity in the blue-detection area *is* the experience of blue (to be precise, the activity in that area plus activity in whatever other areas it connects up with in the right way give rise to a blue experience in you and for you). Different segments of your nervous system being activated in different ways by different color chips is all there is to your color experiences. We can argue forever about whether we have mere correlations or real identities before us. That is, we can both, if we wish, stand our ground.

However, I will claim this. First, my explanation is simpler than your dualistic one, which assumes that all we have is a correlation between nonphysical mental events and physical brain events. Simplicity is to be favored, all else being equal. Second, my explanation is philosophically and scientifically more credible than yours, since you can say nothing sensible about the nature of nonphysical mental events. Third, your position assumes that there is an unbridgeable gap between the subjective and the objective realms. There isn't. Again it is easy to explain why certain brain events are uniquely experienced by you subjectively: Only you are properly hooked up to your own nervous system to have your own experiences.

Subjective Realism and Phenomenal Consciousness

Easing the pain of accepting this picture requires a further piece of therapy. One thing many people fear about a naturalistic view of mind is that they think it will, by identifying mind with brain, make experiences a thing of the past. The worry goes something like this: The Cartesian picture of mind begins (and possibly ends) with recognition that we humans possess first-person phenomenal consciousness, that there is something-it-is-like to be a subject of experience. We are not mere information processors. The scientific picture of mind identifies the mind with certain objective physical processes. But the subjective and the objective can't be meshed or melded. First-person phenomenal consciousness cannot, even in principle, be captured in the sort of third-person objective description that normal science relishes.

There is something right about this point, although it is implausibly used by many philosophers—the ones I dubbed "Mysterians" a decade ago—to argue that although the mind *is* a natural phenomenon we humans are not smart enough to ever grasp or make intelligible its nature. Why's that? Because we have no conceptual resources, nor are they in the offing, to comprehend a phenomenon as both subjective and objective. Certain *scientistic* types, some reductionists and eliminativists, inadvertently encourage mysterianism by seeming to suggest that subjective experience, once seen for what it is, disappears—it is reduced to something else, say the activity of cell assemblies, or eliminated, the way "phlogiston" was eliminated as an explanation of heat exchange.

To have any chance of being accepted by defenders of the manifest image, a naturalistic theory had better provide an account of phenomenal consciousness. It had better show how subjectivity is possible in a material world. And it can. Token physicalism is the view that each and every mental event, each and every experience, is some physical event or other—presumably some central nervous system event. We can accept the truth of token physicalism, and thus reject the Cartesian view that denies it, while resisting the conclusion that the essence of a mental event is captured completely by a description at the neural level. The reason applies uniquely to conscious mental events, which are essentially and uniquely Janus-faced. They have first-person subjective feel *and* they are realized in objective states of affairs. Water and gold would be H_2O and the substance with atomic number 79, respectively, even if there were no subjects in the world. *Objective realism* is true of water and gold. At least it is not crazy to think this. But even if a conscious mental state, say your experience here and now of reading this sentence, is realized in neural event *n*, it is *not* the case that *n* could occur in a world without subjects—in particular in a world without you, since *n* is the way the experience is uniquely realized in you. It is fine with the token physicalist if, for each reader, the neural realizer of the experience of the relevant sentence is somewhat different, so long as there is always a neural realizer.

The objective states of affairs that are conscious mental events are unique in producing first-person feel—*phenomenality*. If certain objective states of affairs obtain then so do first-person feels, and if there are first-person feels then the relevant objective states of affairs obtain. The asymmetry between water and gold on the one side and conscious mental events on the other comes to this: The nature of water and gold is essentially and completely objective, ergo objective realism. The nature of conscious mental events is such that despite being perfectly natural, objective states of affairs, they have as part of their essential nature their subjective feel. Call this basic idea *subjective realism*.

Subjective realism says that the relevant objective state of affairs in a sentient creature produces certain subjective feels for that creature. The subjective feel is produced and realized in an organism by virtue of the relevant objective state of affairs obtaining in that organism. It, the subjective feel, is, as it were, no more than the relevant objective state of affairs obtaining in a creature that feels things. However, since the relevant objective state of affairs is only described or captured as the thing it is, in this case, a conscious mental event, as it is captured or felt by the organism itself, a completely third-person neural description of it doesn't capture it. The reason is that third-person descriptions don't capture feels. Certain third-person states of affairs are the realizations of feels, but the feels are only had or captured by (or in) the creatures in whom those states of affairs obtain.

Thus one may be committed to the truth of physicalism without being committed to the claim that the essence of an experience is captured fully by a description of its neural realizer. It produces a mental cramp to think that mental events are neural events but that their essence cannot be captured completely in neural terms. Such is the power of objective realism, a doctrine that applies to most types of things in the universe but not to experiences. The cramping can be eased, I think, by accepting that the subjective realist is claiming nothing supernatural. It is simply a unique but nonmysterious fact about conscious mental states that they possess a phenomenal side. If you don't mention *that,* and possibly *how,* these states appear first-personally, you have failed to describe one, possibly two, of their essential features. Your metaphysic is incomplete. If you see things in the Janus-way recommended here, the intuition that there is an unbridgeable explanatory gap between conscious mental states and their realizers is deflated. Possibly it disappears.

There is another, related way to make the point in favor of subjective realism. This way turns on paying attention to indexicals, in particular to pronouns. "I" is an essential indexical from the point of view of the subjective realist because it essentially and uniquely captures or, at least, essentially marks the first-person feels that I have been discussing. Description and explanation in normal mind science occur in an objective third-person or impersonal idiom:

> When an individual organism, O, sees a blue cube there is binding of activity in the color and shape sectors of the brain.

Generalizations of this sort do not capture first-person feel. They *assume* that there is first-person feel but they don't capture it. First-person feel is only captured by the subject of experience. And this is why the first-person pronoun is needed to explicitly mark experiences. I say "explicitly mark" because sentences like "Blue cube, here, now" also do the job, but only if it is assumed that the subject is, as it were, marking herself as the site of the "blue-here-now" experience.

The individual gripped by the manifest image, in particular by the Cartesian idea that we possess nonphysical minds, makes this sensible demand on any naturalistic view that could even be entertained as a replacement: Don't mess with phenomenal consciousness. It is a given that I and all my compatriots are subjects of experience. You will need to say more than that a physicalist conception of mind is simpler than a dualist view to capture my interest. All sorts of views are simpler than their opponents—for example, that water is the only element is simpler than every view that countenances more than one element—but they fail because they are miserably simplistic. Simplicity is an interesting feature of a view only when it explains everything that needs explaining. And in the case of mind, the main thing that needs explaining is how there could be phenomenal consciousness in a material world.

Although we have now seen how this can be done, this does not mean enough has been said to win over the Cartesian. The subjective realist is a physicalist who claims that she can meet the plausible demand of the Cartesian to account for—or at least leave ample space for—phenomenal consciousness. For the subjective realist, as for the Cartesian, it is a fundamental fact that phenomenal consciousness exists and is in need of explanation.

It is this sort of thinking that makes it credible to claim that the regulative assumption that mental events are brain events has reached the status of being a constitutive thesis among mind scientists. That is, it was initially assumed but has panned out as highly plausible. It explains everything the Cartesian view can explain, but in a nonmysterious way that fits much better than the Cartesian view into a unified naturalistic picture of the world.

Things are somewhat different with the thesis of causal determinism. This too is a regulative assumption of contemporary mind science. Has it, like the assumption of mind/brain identity, achieved constitutive status? The answer seems to depend mainly on the eye of the beholder. Mind science has so far discovered no causal laws at the level of mentality. The "law of nature" that mental events are brain events is not a causal law in the relevant sense. It is only an expression of the regulative ideal of naturalism of the subjective realist stripe, which keeps panning out, in the sense that it survives without counterexamples or metaphysical incoherence, sufficiently to maintain its regulative status.

But there are no interesting and general laws that tell us in the human case that some set of conditions, C, is necessary and/or sufficient to produce some (type of) thought, T, or some (type of) action, A. Even for a single human act, say my decision to go to the movies last night versus staying home and reading a book, no one is positioned to give a complete causal account of why I chose to go to the movies rather than stay home.[11] Although it is, as I have said, assumed that some finite set of causes exists for why I went to the movies, no one can remotely produce the goods.

In my experience, many people latch onto this fact that mind science lacks strict causal laws and for this reason refuse to grant the assumption of the ubiquity of causation constitutive status. Others, like myself, think the assumption leads to progressive research and survives without counterexamples (no miracles occur) and thus that it deserves the imprimatur of being constitutive. But the current state of play is such that there is something of a standoff here. I'll have much more to say about this matter in the next chapter. But saying this much marks the route we'll be travelling with some clear cautions.

The fact remains that mind scientists, in my experience, assume that strict causal determinism is true. They assume that for each and every effect at the level of mentality, there exists a set of causes that are sufficient to produce that effect and that taken together necessitate that effect. Consider this analogy: Suppose I decide to regulate my economic behavior by assuming that if I do things correctly, I will become very rich. This is probably an idle hope. But by realistically looking for ways to become very rich, I might become richer than if I didn't make this unlikely assumption. I think the strict deterministic assumption performs a similar service for contemporary mind science. Before we can

[11]To be clear, both mind scientists and Cartesians assume that there is a causal account. The first assumes that the proximate causes of my decision are in my brain, the second that they are mental and immaterial. And the first, but not the second, usually assumes that there are causal forces outside my head that taken together with the ones in my head are sufficient to explain what caused my action. The Cartesian assumes that these other "causes" function, at most, as influences, as information I use in making my decision as prime mover unmoved.

even hope to put forward interesting laws stated in terms of necessary and sufficient conditions, we will need to describe the terrain to plot out the mind/brain's causal topography and to explain how brains are connected to the rest of the body and the world.

The case of vision research is telling. Vision is the best-understood area of brain function—but even here there are no strict deterministic laws. The visual system of the rhesus macaque has been analyzed into some 25-plus cortical subsystems, and the complex probabilistic relations among these components are fairly well understood. But the system is so complex that despite understanding pretty well how the visual system works, the task of giving a complete causal account for how exactly any individual perception occurred is something no sensible neuroscientist would even attempt. We can say such things as when a macaque monkey is presented with a case of binocular rivalry in which the stimulus is one of moving horizontal and vertical lines, the visual system will lock on seeing vertical or horizontal lines alternately, but never both at once. We know what sets of neurons are activated in each case. But no one thinks the isolation of the differential activity of such small groups of neurons yields anything like a complete causal explanation of why the monkey sees what it sees. First, everyone acknowledges that there are other areas involved in motion detection than those examined in these experiments. Second, the activation of the "vertical" and "horizontal" neurons involve, for each visual lock, only certain percentages of the relevant regions, and no one understands why 80 percent activation is as good as 100 percent in yielding the result of seeing vertical or horizontal lines. The upshot is simply that this research (like all other brain research I am familiar with), though it is rightly seen by neuroscientists as beautiful and informative work, yields nothing remotely close to a strict deterministic law of visual perception.

So why assume determinism, if even for the best-understood piece of mental machinery all we have is a good functional analysis of how it works? We possess at present no laws stated in terms of necessary and sufficient conditions for "the prototypical" visual system, let alone for the processing of individual visual systems.

The best answer is that one should "keep on truckin'." By assuming the strongest possible type of causal relations, mind scientists, like their brethren in other sciences, take whatever functional regularities they discern and whatever statistical regularities they discover, and try to tighten the causal net by seeking more precise descriptions, analyses, and statistical generalizations. Even if the causal net cannot, even in principle, be tightened to the point of escaping statistical laws, assuming that it can will lead to the discovery of the most complete causal story possible, whatever that story is. So the second regulative assumption of mind science inspires mind science to "keep on truckin'," and in the current situation of vast ignorance about mind that is not a bad thing.

WHAT I HAVE JUST SAID opens up the possibility for another strategy for dealing with the conflict between the manifest and scientific images. This strategy points out that we are thoroughly clueless about how the singularity that banged at the Big Bang got there in the first place, and thus that astronomy provides ample room for the idea of God as creator. Likewise, we have nothing like a complete theory of how the mind works, and thus the idea that the mind has powers of free agency—or what is different, that it is at root a causally indeterministic system—is hardly foreclosed. There is something to this idea. One might think, however, that it is not particularly stable. Even if it is possible at the beginning of this millennium, its roaming space is being diminished, albeit not totally eliminated, each day by advances in astrophysics and mind science. However, I'm not sure if this is a credible response.

Cosmologists do not in fact have much to say about how or why the singularity that banged got there. Some will say this is an ill-formed question because it asks us to talk about the singularity's origin in temporal terms, which is absurd, because there was no such thing as time until the Big Bang occurred. But others think that the question, or at least the impulse behind it, makes sense, and will admit that astrophysics has no resources to address it. In the case of mind science, there is no doubt that we are gaining more and more information about how the mind/brain works with each passing day, and that our knowledge is increasing exponentially. But even if you think, as I do, that we will come to understand the mind/brain very deeply, it is not unreasonable to bet that our understanding—say, in one hundred years—will still be largely framed in terms of statistical generalizations about functional relations among components, subcomponents, and various neurochemicals. We will know the mind/brain in terms of complex statistical relations among the component parts of mobile beings interacting with the world.

If we never discover a neat set of deterministic laws governing the mind/brain, this need not be a scientific failure. Perhaps there are such laws but there are too many variables, in too many combinations and permutations, for the hypothesized strict laws to be feasibly stated. Or perhaps it is the nature of the beast—in this case, us—that whatever laws do obtain are inherently probabilistic. There is no harm in looking for strict deterministic laws or in hoping they will turn up—just as (within moral bounds) there is nothing wrong with my trying to become very rich. If the assumption that the mind is obedient to causal law at every junction continues to lead—as I claim it now does—to progressive research in mind science, and if it survives without counterexample, then it is a wise assumption. Better than the alternative.

WILLIAM JAMES, one of my personal heroes, was as obsessed as any nineteenth-century thinker with how to mesh his scientific and humanistic commitments. In *Psychology: The Briefer Course*, just after he offers the judgment that psychology circa 1890 contains "not a single law in the sense in which physics shows us laws," James writes:

> Let psychology frankly admit that *for her scientific purposes* determinism may be claimed; and no one can find fault. If, then, it turns out later that the claim has only relative purpose, the readjustment can be made. Now ethics makes a counterclaim; and the present writer, for one, has no hesitation in regarding her claim as the stronger, and in assuming that our wills are "free." For him, then, the deterministic assumption of psychology is merely provisional and methodological.

He adds:

> This is no place to argue the ethical point; and I only mention it to show that all these special sciences, marked off for conven-

> ience from the remaining body of truth, must hold their assumptions and results subject to revision in the light of each other's needs. The forum where they hold discussion is called metaphysics. . . . And as soon as one's purpose is the attainment of the maximum of possible insight into the world as a whole, the metaphysical puzzles become the most urgent of all.

James, then, deemed the deterministic assumption of mind science as "merely provisional and methodological" precisely because, knowing as well as anyone on earth the actual achievements of scientific psychology, he saw that it contained not a single deterministic law. But why did James—why would anyone—approve of a regulative assumption that he didn't believe would pan out? Why does he say that "no one can find fault" with the deterministic assumption of mind science, given that he thinks it cannot be abided by ethics? The only answer I can think of has to do with James's pragmatism. The mind scientist has one and only one aim: to get mind science up and running and on the road to maturity. Guiding his inquiry by the strong assumption that mind science will discover laws of the same strictness and generality that physics, chemistry, and astronomy possess, will yield as much and as deep knowledge of the lawlike generalizations pertaining to the mind/brain as can possibly be attained. If one expected only loose and vague generalizations, then one might rest satisfied with less than is there. Assuming strict causal determinism will lead, better than any weaker assumption, to the discovery of whatever regularities, of whatever strength, the mind abides.

There is still one odd thing about James's formulation. He doesn't say (as I would like him to) that we have, on the one side, an underdeveloped and inchoate metaphysic of morals within which the idea of "free will," whatever exactly it is, has a place. And that we have on the other side the beginnings of a science that seeks to understand human minds, and ultimately persons, scientifically. If he had said this, then his conclusion that we wait and see how things work out would make sense. That is, we recognize that there *appears* to be a conflict between the assumptions of mind science and the assumption of ethics that we possess free will, and thus between the scientific and manifest images. But we take the stance of the patient pragmatist:

Wait and see, if as mind science progresses, whether and how it accounts for the fact that humans are agents who reason, deliberate, and choose. When the right time comes, as the ethical conception of humans as free agents and the scientific image of humans as complex animals both develop, bring them together into the court of metaphysics to see whether and how human agency is accounted for, and see, in particular, if the apparent conflict is real.

The Manifest Image, Again

So here we have these two assumptions: (1) that the mind is the brain, or better, that mental events are brain events, or better still, that mental events are central nervous system events or even that they are whole-person events; and (2) that the mind is obedient to natural law, possibly strict causal ones. These two assumptions regulate modern mind science but offend against the dominant manifest picture of persons.

The manifest image, as we've seen, conceives the mind and mental causation in precisely opposite terms. The mind is incorporeal. Furthermore, when I choose to perform some action, I do so of my own free will. That is, I choose to act and then act in a way that is not subject to ordinary causal law.[12]

---

[12]Most people, most Americans at any rate, believe that they have souls, that there is an afterlife, and that they have free will, and they believe in angels and in God. CNN, *USA Today*, and *The New York Times* report that as many as 90 percent of Americans believe in these things. To be specific, according to the most up-to-date statistics I have seen concerning belief in God, heaven, and hell (Natalie Angier, "Confessions of a Lonely Atheist," *New York Times*, January 14, 2001), 92–97 percent of Americans claim to believe in God, only 3 percent claim to be atheists, or true disbelievers (in some European countries, Holland, Sweden, France, and the Czech Republic, for example, atheists comprise between 15 and 20 percent of the population); 86 percent believe in heaven; and 76 percent believe in hell. Time and again, belief in souls, angels, and free will closely approximate the percentage of believers in God. And this is not remotely surprising since these ideas are mutually supported within the perennial philosophy that refines and endorses the dominant humanistic image. According to most polls, 60 percent of scientists claim to believe in God. And a full 40 percent of scientists listed in *American Men and Women in Science* not only believe in a personal God, but that he listens to their prayers. Interestingly, among members of the elite National Academy of Sciences only 7 percent believe in a personal God.

Beyond this basic conflict, I've also explained in this chapter that the two opposing pairs of assumptions have somewhat different status. The idea that our minds are not incorporeal but physical—that they are the kinds of minds to be expected in embodied social animals that have evolved according to Darwinian evolutionary principles—has reached something like the status of truth within mind science. This assumption has moved, as it were, from the status of a hypothesis, a pure regulative assumption, to the status of a constitutive truth—again within mind science, but not within the manifest image. But perhaps it can be absorbed into the manifest image once it is better understood that the scientific conception of the mind/brain does not undermine the idea that we are subjects of experience, possessed of phenomenal consciousness.

Free will is disbelieved within mind science. At least, mind scientists operate as if everything that happens, including how we think, choose, and act, is governed by the principle of universal causation. This regulative assumption has proved fruitful, but, unlike the physicality of the mind, it has not been shown to be *more* than useful.

This difference in status, once pointed out, can be readily seen in everyday mind science. For instance, thanks to brain imaging techniques, we can actually point to many brain processes that are plausible candidates for being certain experiences. Such results warrant believing that mental events, at least these ones, are brain processes, or better, central nervous system processes. But there is not one case of a human thought or action for which anyone has been able to cite, list, or compile the complete set of causes that produced it.[13] No one is even trying to assemble such a list.

With this in mind, I want now to explore more deeply the status of the assumption that we are possessed of Cartesian free will. Despite the lack of even one completely adequate counterexample, I think it is an incoherent idea.

---

[13]One might claim, however, that something close to complete explanations are (almost) available for some simple sensations and visual perceptions.

FOUR

Free Will

> "You sound to me as though you don't believe in free will," said Billy Pilgrim.
> "If I hadn't spent so much time studying Earthlings," said the Tralfamadorian, "I wouldn't have any idea what was meant by free will. I've visited thirty-one inhabited planets in the universe, and I have studied reports on one hundred more. Only on Earth is there any talk of free will."
>
> —*Kurt Vonnegut*,
> **Slaughterhouse Five**, *1969*

WHAT IS THE DIFFERENCE BETWEEN a wink and a blink? The answer, we all know, is that winking is voluntary, we do it for a purpose. Blinking is involuntary. It also has a purpose, keeping our eyes moist, but we don't will that it be done. The way we are wired makes sure that blinking gets done.

Winks mean something. Depending on the context, they can mean "You're cute," "Yeah, he does tend to go on and on," "Don't sweat it," and many other things. But blinks don't mean anything at all.

A theory of mind must account for the difference between winking and blinking. We have every expectation that it will do so by citing the role of intentions and motives in winking but not in blinking, and that it will preserve the idea that winking is something we do, whereas blinking is something that happens.

But is the distinction really so clear? Many neurophilosophers and mind scientists nowadays describe the mind as a massively parallel processing system. Beneath consciousness a "parallel pandemonium" operates. Consider Hans Zinsser's account of the process of choosing the mayor in Hurdenberg, Sweden:

> [I]n Hurdenberg in Sweden, . . . in the Middle Ages a mayor was elected in the following manner. The persons eligible sat around a table, with their heads bowed forward, allowing their beards to rest on the table. A louse was then put in the middle of the table. The one into whose beard it first adventured was the mayor for the ensuing year.

Usually we think of the choice of a mayor as a rational matter. In some places he or she is appointed on his or her merits, in others he is elected by a majority that judges the winner more meritorious than the other candidates. But suppose the "lousey picture" is a metaphor for an internal process. When the mayor, with powers to appoint his successor, deliberates about the worthy candidates, he lines the candidates up in imagination and considers the pros and cons of each. Assume there are some close calls so that at any time before the choice is made, several candidates remain viable. How does he choose among them? The standard view is that the deliberative mind gets down to nitty-gritty details and produces a ranking, at which point it decides. But suppose that what really happens is that the neural equivalent of a louse just jumps onto the mental representation of one candidate, which causes the mayor to choose him. Alternatively, perhaps deliberation does get more and more subtle and fine-grained, but it does so not by mustering good reasons pro and con, but because it is forced into seeing the outstanding merits of one candidate over the others because of something like a louse gaining control of the deliberative circuitry.

We would not like it if this were how things worked. It would undermine our confidence that we deliberate rationally about matters of importance. But the "lousey picture" is out there and we will need to deal with it.

Suppose I wink at you in order to express the thought that I think you are cute. You respond by slapping me in the face or leaving the room, because you find my wink rude and lascivious. I will now wish I hadn't winked.

But suppose I think of my actions along lousey lines. I apologize for having offended you but explain that what I did was inevitable. My winking circuitry is on a direct upstream link from my cuteness-assessing circuitry, which your presence caused to reach the right threshold to cause me to wink. Similarly my apologizing circuitry is now being activated because it gets activated when people take offense at something I've done.

This, of course, is not the usual way we think about such episodes. How do we normally think about them? One alternative is in Roderick Chisholm's way. I see you, judge you cute, and decide to wink to convey this thought. When I decide to wink and then do so, I exercise "a prerogative which some would attribute only to God: each of us when we act, is a prime mover unmoved. In doing what we do, we cause certain things to happen, and nothing—or no one—causes us to cause those events to happen." In other words, the muscles of my face move in a certain way because I will it without any set of causal antecedents determining the state of my will.

Now this is not the way we normally think about such episodes either. To think that my winking at you involves exercising a God-like prerogative, according to which I was the sole originating cause of my winking, seems preposterous. To be sure, I wondered whether winking would be well received. I considered several other ways of getting your attention. I now see that it was not the best idea.

But if asked to explain my action I might say that (a) I find you attractive, (b) I have seen winking work well for Humphrey Bogart, Brad Pitt, and Richard Gere in movies with the likes of Ingrid Bergman, Julia Roberts, and Meg Ryan, and (c) I was raised among men who think winking to express "you're cute" is okay and among women who don't find it offensive. To be sure, I take responsibility for choosing winking over the other options, and I really am sorry I offended you. "Now, will you have dinner with me?"

I claim that no sensible person would argue that the reasons I just gave were *not* components of a complex causal process that eventuated in my decision to try my ill-fated wink.

If this is right, then our normal way of thinking about an act such as a wink accords neither with the lousey picture, in which a wink is really just like a blink, nor with a picture in which we exercise God-like creative abilities, *ex nihilo, ab initio, de novo*—or whatever other Latin words convey the underlying implausible thought. Philosophers call this picture "agent causation." When an agent judges or acts freely, she initiates the judgment or action and nothing and no one causes her to do so.

The trouble is that even if, in the normal course of things, we do not speak as if we are the sole, originating cause of our voluntary actions, perennial philosophy presses the manifest image to think that if we possess free will, it must be something like what Chisholm says it is, a form of agent causation.

If this is right, Descartes's historical role takes on new meaning. When I introduced the Cartesian conception of mind as an incorporeal substance, I claimed that Descartes played the role Sellars claims the perennial philosopher normally plays, refining and endorsing the manifest image of mind according to which the mind was nonphysical and distinct from the body. But I don't think it was inevitable that the manifest image had to fall into the grip of the idea that the mind is nonphysical. For myself, if I were asked to say how my consciousness *seems*, I would say that it doesn't seem either physical or nonphysical. It is not as if it seems some third way. It just seems like what it is—consciousness.[1]

One reason, however, that the belief in a nonphysical mind took hold is because thinking of the mind as nonphysical fits well with thinking of human agents as free. Physical things obey natural laws, nonphysical things don't. Or so one might plausibly think.

But whereas Cartesian dualism resonates almost perfectly with the manifest image as it has evolved in the West, it is less obvious that the Cartesian picture of free agency does so. In my experience, most people will in fact say that their consciousness seems nonphysical (although you can dislodge that thought a fair amount by asking people to focus on their emotions). The manifest image says this, and it

[1]Descartes is plausibly read as claiming to experience his consciousness as nonphysical—thinking *is* unextended. William James, however, claims to experience his consciousness as some sort of biological process. "Whenever I try to become sensible of my thinking activity as such, what I catch is some bodily fact, an impression coming from my throat or head, or throat, or nose." Edmund Husserl experienced consciousness in Descartes's way, whereas Maurice Merleau-Ponty experienced it James's way. In each of these competing cases, one might wonder to what degree the report of how consciousness seems is a pure untutored phenomenological report, as opposed to thinking or reflecting on consciousness from a certain theoretical perspective.

has infiltrated the hearts and minds of most Westerners. But I do not think most people will say, from an ordinary unreflective pose, that their wills are free in Descartes's and Chisholm's strong sense.

Descartes says that "the will is so free in its nature, that it can never be constrained. . . . And the whole action of the soul consists in this, that solely because it desires something, it causes a little gland to which it is closely united to move in a way requisite to produce the effect which relates to this desire." Chisholm's idea that "each of us when we act, is a prime mover unmoved" simply repeats Descartes's picture of the will as totally unconstrained. But if what I said above about winking is true, we don't think about voluntary actions in either way. The lousey picture of winking is ridiculously deflationary. It dissolves the distinction between involuntary and voluntary actions. But the unconstrained God-like picture is absurdly inflationary. It puffs up the will to an unbelievable degree. When we explain our voluntary actions, we never speak as if these acts were totally unconstrained. We think of free actions as unforced, as not involving compulsion to do what we don't want to do, but that is very different from thinking that free actions occur with no constraints or totally outside the causal nexus.

But if that is true, then I must perform a different sort of philosophical therapy from what I did in the last chapter. With regard to the Cartesian picture of mind, I claimed that one way to diminish the threat of the scientific image is to see that it not only does not threaten, but in fact preserves the Cartesian idea that we are conscious agents even if consciousness is not realized in an incorporeal soul. The question arises: What threat do we need to save free will from given that our ordinary commonsense way of speaking about it does not seem to conform to the industrial-strength Cartesian view?

One possibility is that the scientific image's threat to free will can only be seen on a distinctively philosophical level. Most people think that free actions are produced within complex causal contexts and by complex causal processes that include, as a central feature, our will. We typically speak of all sorts of considerations and influences that went into some decision. No one speaks as if they made a decision in a completely unconstrained way.

So the free will worry arises only because philosophers, or the philosophically minded, press the following disturbing thought: "Science says everything is determined. Determinism is incompatible with free will. You speak as if your agency works this way: You are affected by all manner of influences, your genes, your upbringing, the options available to you, and so on. And then you think you use your free will to decide what to do. But science doesn't allow that. Why? Because if everything is determined, then so is the state of your will."

"There is a way out. Conceive of your will along the lines suggested by Descartes and Chisholm, that is, as operating outside the constraints of the world as science conceives it. For free will to exist at all, some part of you must operate, however many worldly constraints and influences you face, as a prime mover itself unmoved, unconstrained by anything."

If this diagnosis is correct, then it is not quite right to say that free will is a major locus of conflict between the manifest and scientific images. Rather, worries about free will are generated among ordinary people by perennial philosophy's stance that the only logical way of conceiving of free will is, upon reflection, incompatible with science.

If this is right, our therapeutic route will need to be more circuitous than it has been so far. We will need to ask: Is it true that determinism is incompatible with free will? Or, what is different, is determinism incompatible with the idea of voluntary thought and action, with freedom of thought and action, with rational behavior?

My answer to these questions is the same. It is "no." Those who abide the humanistic image therefore needn't catch the dis-ease fostered by perennial philosophy. I'll explain, but it will take time. It will be useful, as a way of keeping track of what we in fact want to gain from any acceptable account of the voluntary, to make a list or scorecard. Here are the attributes of free will that we value:

- Self-control
- Self-expression
- Individuality
- Reasons-sensitivity
- Rational deliberation
- Rational accountability
- Moral accountability
- The capacity to do otherwise

- Unpredictability
- Political freedom

I will argue here that we can have—indeed do have—all of these things without needing Cartesian free will.

Our Manifest Image Is Philosophically Diseased

The distinction between voluntary and involuntary thought and action is all we should need or want. The concept of voluntary action, assuming it can be made sense of in naturalistic terms, is enough to preserve the idea that some thought and action are free. The added ingredients that are thought to be required to make voluntary reasoning and action truly free—freer than they would be if they were simply voluntary—are rooted, I think, in adherence to a certain theological picture. To put it another way, the part of the manifest image that distinguishes between voluntary and involuntary action is sensible and can be preserved in the scientific image. The addition of "free will" to our conception of the voluntary is an instance where a bad philosophical idea infiltrated ordinary common sense. It caused the manifest image to catch a distinctively "philosophical disease," which, when activated, leads us to believe our free will is being threatened.

In broad strokes, what happened was this: God was conceived in the West as the Prime Mover Himself Unmoved. The Old Testament tells the basic story in Genesis. Aristotle put forward the idea independently from a pagan perspective, and Aquinas, most famously, refined and endorsed it in a way still widely accepted within the perennial philosophy rooted in Judeo-Christianity.

God created the universe without anything causing him to do so. Humans were created by God in his image, but he left our fates up to us. If God is not to be incomprehensibly cruel and arbitrary in how he doles out eternal reward or punishment, humans must possess free wills suitable to finite creatures made in his Image. Without them, it makes no sense to think of persons actually making the kind of choices that would merit their going to heaven or hell. An alternative would be to think that some people luck out by being born into good, secure families, acquire decent characters, and do the right things. Others are born into rotten families and become rotten people. The theological picture refused to tolerate this plausible idea and squashed it like a bug. Humans must have the right kind of will to overcome any luck in life circumstances. What kind of will fits this job description? You guessed it: a capacity to initiate actions where one is not caused to act by any prior set of causes—a will so powerful that people can be judged personally responsible for failure to overcome unimaginably bad life circumstances.

Placing the troublesome libertarian concept of free will in the theological context from which it emerged makes Chisholm's description of the capacity easier to understand and at the same time chillingly otherworldly. Recall again what Chisholm writes: "If we are responsible . . . then we have a prerogative which some would attribute only to God: each of us when we act, is a prime mover unmoved. In doing what we do, we cause certain things to happen, and nothing—or no one—causes us to cause those events to happen."

From this philosopher's armchair this idea looks patently crazy. Furthermore, it is built on a sand castle. The picture of God and his Will from which the Western conception of human free will is inherited by analogy is based on a huge epistemic stretch. That God exists is a matter of great contention. What God's nature is, is a matter of additional contention. That is, even supposing one makes the leap of faith to explain the universe in terms of a divinity, no attributes of the divinity are specified. The creator might be personlike or not; it might be possessed of every perfection, just a few, or none.

Even if one finds the story of the universe as originating with the Big Bang unsatisfying and asks, "Well, how did the singularity that banged get there?" I might answer that I haven't a clue, nor do you. I might say that the singularity was always there (in exactly the same sense that theists say God always was there), or perhaps, grasping for straws, I will say the singularity that banged was born of the Mother of the singularity, which had its own Mother, and so on. Mothers of singularities backward for all eternity. It is not clear that any of these responses is more incoherent than the answer that confidently grounds being in a divinity that was always there, and then has the chutzpah to confidently specify the divinity's attributes.

The troublesome picture of human free will thus arises from a prior troublesome picture of divine agency. There is nothing mandatory about believing in God. Even if one does believe in God, there is nothing mandatory about thinking of God as personlike, as having created humans in his image, having created heaven and hell, and rewarding and punishing people according to the standard story. Only if one accepts this whole picture is one cornered into thinking humans have wills of the sort dear to the hearts of theologians, and a few philosophers, most of them long dead, who bought into this theological picture.

It is unimaginable to me, despite the power of the phenomenological feeling that we are agents who control what we do, that anything as strong as a conception of ourselves as finite unmoved movers would have been added to our manifest image unless we had first conceived of God and his will along these lines, and then added the view that he holds people fully accountable for what they do.[2]

Even if you accept this diagnosis of how we became heirs to a truly bad idea—a dangerous theological germ that took hold and spread among sensible souls whose mental immune systems were down—we need to explain why it continues to cause consternation among atheists, agnostics, and secular humanists. My answer, which is not intended to be glib, is that philosophical diseases, like many others, can have long-lasting effects. Assuming, for instance, one survives the active stage of hepatitis, the liver is still damaged, and one may well suffer cirrhosis, liver cancer, or metabolic problems later.

---

[2]Buddhism is quietistic about creation, and thus it is what we would call theologically agnostic. When Buddhists talk of human agency, they frame it in terms of what they call codependent causation. Roughly, the idea is that everything is in causal relation with everything else within its spatial and temporal bandwidth. Humans cause things to happen and are caused to make things happen. Furthermore, all of these causes and effects produce other, novel cause and effect relations. I both have and produce bad *karma* if I am caused to take the wrong road or if I take myself down the wrong road. Dharmic practices, meditation, reflection, and compassionate acts provide techniques to move myself onto the track of the good.

The bad idea of free will infected our moral and legal conceptions while it was maximally influential. It lodged in the manifest image and resides in the vicinity of our conception of voluntary action. It is now activated to some degree whenever that concept is activated. It can, I think, be excised, but the operation is delicate.

The Current Situation

Almost every philosopher thinks the distinction between voluntary and involuntary actions, between "mere bodily movements" on one side and intentional actions on the other, can and must be preserved. However, many philosophers doubt that its close relation, the concept of free will, can be preserved if the scientific image is true.

You will not be surprised to know that, having been raised a Roman Catholic, I considered these matters pretty important. Sins were bad. There were mortal and venial sins, and a single unconfessed mortal sin could land you a straight ride to hell. Sins, of course, were voluntary. One memorable feature of my own "birds and bees" lesson involved Dad explaining that wet dreams were involuntary and thus neither good nor bad, but masturbation, being voluntary, was a mortal sin and could get you a permanent pass to the place with the hot oil.

In the late 1970s I defended my dissertation, which was on the question of how agency could be understood if one took a behaviorist approach to human action. The first chapter of my thesis was a history of the concept of the "reflex" in philosophy of mind and physiological psychology. The general consensus of people who worked on reflexes was that a wink differed from a blink in that the former was routed through the cerebral cortex whereas reflexes involved subcortical circuits. Winks involved the higher reaches of the central nervous system, whereas blinks, like circulation, respiration, and digestion, involved the autonomic nervous system and originated in the brain stem. A few eighteenth- and nineteenth-century scientists thought that a soul must be involved in any and all movement, and some went so far as to posit a "spinal soul" to explain the movements of decapitated salamanders (as well as the proverbial chicken with its head cut off). But for the most part, the voluntary–involuntary distinction fell pretty neatly and effortlessly to the CNS–ANS distinction.

In 1974, B. F. Skinner published *About Behaviorism*, the sequel to his 1971 book *Beyond Freedom and Dignity*. Fred knew that I was working on the problem of agency and was kind enough to give me a preprint of the book in 1973, and to talk over the issue with me. As any reader of both books knows, Skinner thought the concept of free will was outmoded and dangerous (I agreed with him about free will, but not about "freedom" and "dignity"). He did, however, think that his radical behaviorism had the resources to preserve the distinction between involuntary and voluntary behavior. Involuntary behavior consisted of innate and conditioned reflexes (for example, blinks, and learning to salivate to a bell which had been paired with food presentations), whereas voluntary behavior, such as winking, writing poetry, and doing philosophy, was controlled by learning what to do and how to do certain things in order to get reinforced. The first class of behavior Skinner called "respondent behavior" and the second he called "operant behavior."

Walden Two, which Fred later told me was his favorite among his own books, describes a utopia in which everyone is happy and works for the common good. The so-called contingencies of reinforcement have been carefully engineered to make the members of the community do what is good for themselves and others. They make hammocks that they then market, they tend gardens and children with care, they read and think and discuss issues of common concern. They do all of these things voluntarily, not because they are exercising their free wills but because doing so produces reliable rewards. No one holds a gun to anyone's head. The "contingencies of reinforcement" are built into the environment. People, as animals who like rewards, conform their behavior accordingly: They choose to do the things that are rewarded. When reinforcers are abundant and the means to gain what is good are open to everyone, people are content and flourish. At least this was pretty much the whole story according to Skinner.[3]

[3]Given that people will conform their behavior to positive reinforcement, there is a need to make sure the behavior being reinforced is good—morally good, productive of happi-

What Is Voluntary Is Not Free

The distinction between winks and blinks, as well as Skinner's distinction between respondent and operant behavior, seemed to preserve the distinction between involuntary and voluntary action. But Skinner, as I have said, favored moving the concept of free will from the desktop to the trash. Many think you can't have one without the other, indeed that the concept of the "voluntary" is coextensive with the concept of "free will."

The argument that this is *not* so goes as follows. First, the concept of free will postulates a distinct faculty in the mind—where the mind is conceived along Cartesian lines, as a soul. Psychology has no room for Cartesian minds or souls and thus no room for any faculties unique to such a mind. Second, even if we naturalize the mind and think of it as the brain, there is no evidence that the brain possesses a separate and distinct faculty called "will." Third, when people talk of the will as being "free" from a philosophically endorsed reflective pose (but not when they think about willing commonsensically), they think of it as operating outside the realm of causation. Free actions are self-caused in the strong sense advocated by Descartes and Chisholm. Mind science, as a proper part of science generally, normally assumes that everything that happens has causes sufficient to produce it. Although there are actions and movements, things we do and things that just happen to us or in us, there are no "free actions" as these things are conceived along Cartesian lines. The difference between involuntary and voluntary acts has to do with *how* they are caused; it is not that one sort of act is caused and the other uncaused, or that one sort of act is caused by an agent who chooses in accordance with her "free will."

Most people of ordinary common sense at least verbally equate voluntary actions with actions done of one's free will, saying "I did such and such of my own free will" when they mean they did it voluntarily. In most settings, all we care about is the voluntary–

---

ness and flourishing. Skinner's thinking about ends involved a sort of social Darwinism. The behaviors that are conducive to goodness and flourishing evolve over time by a process resembling natural selection.

involuntary distinction. Philosophers, scientists, and theologians, however, don't equate the two. The widely shared view among them is that to say that an action is free, or done of one's own free will, is to say something more than that it is voluntary. It is to say something philosophically contentious. Theologians by and large adhere to the idea that humans are possessed of free will. Many philosophers and most scientists, if forced to reflect on the matter, reject the contentious version of free will, but they treat it as a real problem. Indeed, the philosophers take the problem very seriously and are even obsessed with it. This may be a mistake. Some ideas aren't worth taking seriously.

But this one is worth my taking seriously, at least, because the belief in free will is a central component of the dominant humanistic image in the West. Thanks to the insistence of perennial philosophy, nothing less than morals and the meaning of life turn on there being genuine freedom of the will.

So some form of therapy is required. Even if the standard picture of free will is incoherent and thus not worth discussing, the fact is that in the West, ordinary people of common sense have wedded this incoherent picture of free will to the ordinary conception of voluntary action. Thus they think, and many philosophers encourage them to think, that science, given the chance, will rob them of their free will.

Aristotle and Dewey

Aristotle championed the voluntary–involuntary distinction long before there was a conflict between the manifest and scientific images. John Dewey, living 2,300 years after Aristotle, was an extraordinarily energetic and thoughtful academic and public philosopher. Along with William James and James's unjustly neglected student Mary Caulkins, Dewey was one of only three people ever to serve as president of both the American Philosophical Association and the American Psychological Association. And like James and Caulkins, Dewey was well aware of the conflict brewing between the humanistic image of persons and the assumptions made by the new scientific psychology. Much of his work was devoted to working out the conflict between the two images.

In an early paper, "The Ego as a Cause," written in 1894 when he was 40 years old, Dewey claimed that the main question facing the science of the mind was whether we can "carry back our analysis to scientific conditions, or must we stop at a given point because we have come upon a force of an entirely different order—an independent ego as an entity in itself?" His answer was that the myth of a completely self-initiating ego, an unmoved but self-moving will, was simply a fiction motivated by our ignorance of the causes of human behavior. He saw no need for the notion of a metaphysically unconstrained will or of an independent ego as an unconstrained primal cause in order to have a robust conception of free agency. For there to be agency we need the ego as a cause, possibly even the proximate cause of what we do. But the ego may serve as the proximate cause of action and still itself be part of the causal nexus.

In 1922, Dewey wrote this: "What men have esteemed and fought for in the name of liberty is varied and complex—but certainly it has never been metaphysical freedom of the will." He's right. Cries for "freedom" are typically pleas for life, liberty, and the pursuit of happiness—that is for political freedom, not metaphysical freedom. Anybody who listens to lots of reggae, as I do, can tell you that Bob Marley, Ziggy Marley, Jimmy Cliff, and Peter Tosh are not calling for metaphysical freedom but for equal rights and justice.

This suggests that the manifest image of an age may tenaciously defend an idea that is not only false but unnecessary. Remember that Pope John Paul II thinks that God created the universe, but that one need not believe literally in the Genesis story of creation to maintain that belief. God, being God, could have created a world that works according to Darwinian principles.

Perhaps the same applies to free will. Many people think they need a notion of free agency that involves a self-initiating ego in order to undergird the idea that they are free. Maybe something else can do the job, something where the distinction between voluntary and involuntary does not turn on a distinction between acts initiated by a completely self-initiating will and those that are fully explicable in causal terms.

In the fourth century B.C.E., in the *Nicomachean Ethics,* Aristotle drew the involuntary–voluntary distinction this way: "What is involuntary is what is forced or is caused by ignorance. What is voluntary seems to be what has its origins in the agent himself when he knows the particulars that the action consists in." What Aristotle had in mind was something like this: An action is involuntary if it results from some sort of compulsion against which effort and thinking are impotent, or if the agent in no way knows or grasps what he is doing.

Notice that the law diminishes the degree to which a person is culpable for a crime in accordance with Aristotle's formulation. If an individual is ignorant of the difference between right and wrong, or compelled to do what she does by insanity, a brain tumor, or a gun held to her head, she is not considered legally responsible for her act.

Or take a nonnormative case such as the way the pupil contracts in response to light. Even if I know the rules governing pupil contractions and will that my pupils not contract to light, it won't work. Pupils are not governed by will. The pupil-contracting system is, as psychologists say, "cognitively impenetrable."

Voluntary action, on the other hand, involves the agent knowing what action she is performing and acting from reasons and desires that are her own. To stick with the same example, if I don't want my pupils to contract there is something I can do, namely close my eyes. If I know what is happening and can find a system or subsystem that is cognitively penetrable, in this case the motor system, I can intervene to get the result I want. This involves self-control, the first item on the preceding scorecard.

Rational deliberation, rational accountability, and reasons-sensitivity (three other items on the scorecard) come into the picture in the following straightforward way. When I act voluntarily, I act from reasons, and I can be asked to explain them. Normally, I do so by explaining and defending my reasoning. Even where my voluntary action has minimal effect on any other agent, I may be asked to explain myself. Suppose I get an algebra problem wrong. My teacher, who aims to help me do better, asks me why in this step I thought such and such a move made sense. I explain why I did what I did, why it seemed like a good idea. I give an account of my reasoning. I am told how what I thought led me down the wrong path. This, perhaps, enables me to reason better the next time.

Aristotle is silent about the source of the capacities to think and act with reason—although he does say that "what is voluntary *seems* to be what has its origins in the agent himself." But this silence should not be taken for neutrality on the existence of natural causal origins for these capacities. For Aristotle, and for many post-Cartesian thinkers such as David Hume, John Stuart Mill, and John Dewey, the assumption is that the capacities that are deployed in the initiation of voluntary action are distinctive, but perfectly natural, human capacities with perfectly ordinary natural histories. That the will *seems* to be self-initiating is an understandable illusion. We are not in touch, in a first-person sense, with most of the causal factors that contribute to who we are and to what we do. It is hardly surprising that we are prone to overrate the causes we are in touch with first-personally. When I deliberate and choose among the options before me, I am in touch with the relevant processes, the processes of deliberation and choice. I am not in touch with—indeed I am normally clueless about—what causes me to deliberate and weight my options as I do. So I make a misstep and think deliberation is self-caused. It *seems* that way, after all.[4]

But perhaps, as with creation and Darwin, there is room for accommodation on the matter of free will. The option I have in mind is not one the pope would find congenial. But, that aside, the basic idea is this: Accept that deliberation and will exist and that they are often proximate causes of behavior. Concede, however, that they themselves are natural phenomena, indeed that they are brain processes—subject to whatever causal laws govern proximate brain causation. So deliberation and volition exist. They are not the mental mates of phlogiston, the ether, Santa Claus, or the tooth fairy. But how they *seem* is not, in all respects, how they *are*.

[4]That persons are in first-person touch with high-level proximate causes is a brilliant piece of evolutionary design. Were we in touch with *all* the causes of thought and action, the mind would be much too noisy, and it is doubtful that a noisy mind would enhance fitness. So a design that puts us in touch with proximate causes and screens off distal causes is fitness enhancing. But it has one unfortunate consequence. It causes us to overrate proximate causes and to think that how things seem from the first-person point of view is how they *are*.

The position that the reality of voluntary thought and action is compatible with an analysis of such action *as caused* is called, not surprisingly, *compatibilism*. Those who, like William James, would disparage it as a form of sleight-of-hand call it "soft determinism." For reasons that will become clear, I claim not to be a compatibilist but a *neo-compatibilist.*

In any case, for the strategy to work we must show that the ingredients necessary for free action are compatible with causation. What are those ingredients? They are such things as the ability to pay attention, the causal efficacy of conscious deliberation, reasons-sensitivity, the capacity to act in accordance with desires, and the capacity to consciously monitor and guide action—all items on the scorecard or necessary components of them.

The fear is that a natural creature, an organism subject to natural law—to causation—lacks these attributes. But that is the big mistake. First of all, far from ruling out causation, the concepts of control and self-control *are* causal notions. This is vivid in Daniel Dennett's analyses of the meanings of "control" and "self-control." According to Dennett, we can define control as follows:

> CONTROL: A *controls* B if and only if the relation between A and B is such that A can *drive* B into whichever of B's normal range of states A *wants* B to be in.

This definition captures most of what we mean when we say "A controls B" when A and B are separate entities—when, for example, A is a person and B is A's car.

What about when B is a part of A? Suppose I want to attach a tabletop to table legs with screws. A is me with an active state of my will that contains both my desire and my strategy. B is my motor system. I want to screw in the screws, and this requires that my desire, my decision to do so, activate my motor system, B, in a certain way. This involves self-control, which Dennett defines as follows:

> SELF-CONTROL: For some integrated system S, some *subsystem* S_a *controls subsystems* $S_1 \ldots S_n$ if the relation between S_a and S_1

. . . S_n is such that S_a can *drive* S_1 . . . S_n into the states S_a *wants* them to be in.

These concepts of control and self-control (scorecard items, again) are silent about how the controlling mechanisms' causal powers originate and what exact mechanisms subserve the control relation. But they are thoroughly compatible with there being some such causal antecedents, both ontogenetic and phylogenetic, as well as specific mechanisms governing the activation of (in the present case) the motor system by desire, deliberation, and choice. Indeed, the philosophical naturalist thinks it is mind science's job to fill out the causal picture on both matters.

But the basic idea is that there is a robust conception of free agency that does not require us to be metaphysically free. The picture is this: Persons have free agency so long as the mind has the following architecture. Genes and life experiences feed into a brain that has, as one of its properties, the capacity to process and access information consciously or subconsciously in a way that is one important contributor to, possibly the proximate cause of, a decision, as shown in Fig. 4.1. Recall that the picture of agent causation that posits a free will itself unmoved looks like the diagram in Fig. 4.2.

There is a third view, entertained by some contemporary mind scientists, called *epiphenomenalism*. Epiphenomenalism says that the existence of conscious deliberation is not evidence that this deliberation is important in the causation of action. Roughly, the idea is that

FIGURE 4.1 A Natural Mind

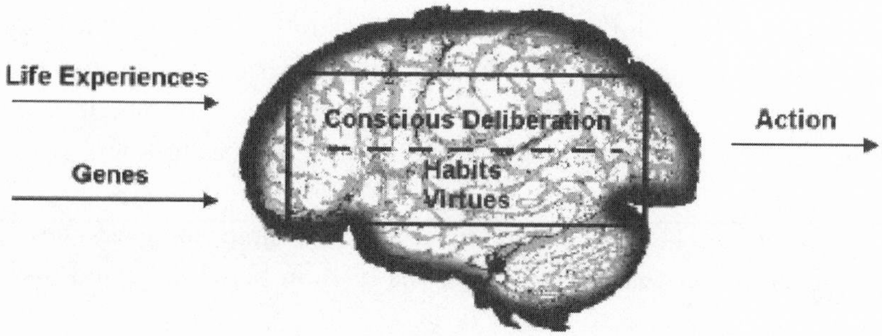

Life Experiences
Genes
Conscious Deliberation
Habits
Virtues
Action

conscious deliberation or choice is itself a side effect, or aftereffect, of the relevant causal processes, in the way sizzling is a side effect of frying eggs or, if you accept the James–Lange theory of emotions, fearing the bear comes after you run. The fear doesn't cause you to run. The epiphenomenalist picture looks like Fig. 4.3.

Now one of the main things we want any acceptable account of voluntary thought and action to do is to preserve the idea that my thoughts and actions are self-expressive. When I do what I choose to do, I am expressing myself, my personality, my character. To put the matter slightly differently, when I perform a voluntary action I intend

FIGURE 4.2 Agent Causation

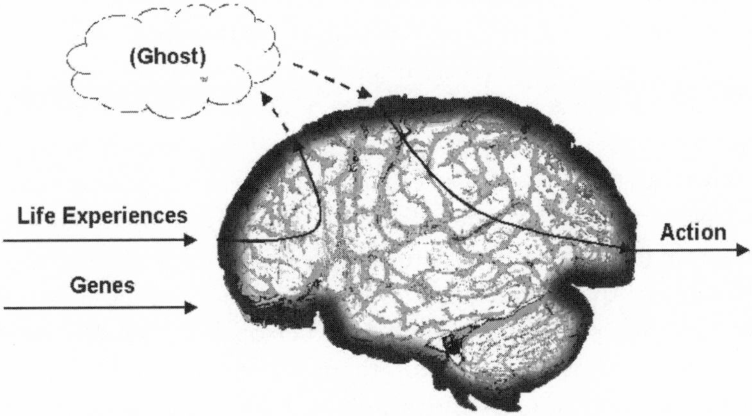

(Ghost)
Life Experiences
Genes
Action

FIGURE 4.3 Epiphenomenalism

Feeling of
Conscious Willing
Life Experiences
Genes
Action

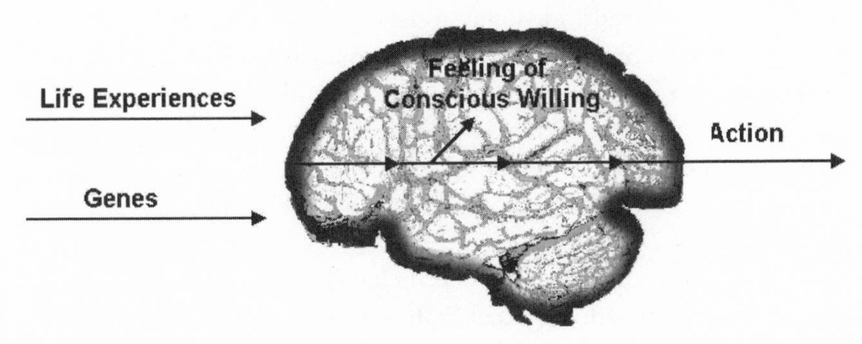

to do what I want, and I want my wants to be expressive of who I am (scorecard item). Furthermore, I want whatever conscious deliberation is involved in deciding to do something to have an actual role in bringing about what I do.

If epiphenomenalism were true in general, then this way of conceiving of voluntary action wouldn't work. In the epiphenomenalist's picture, conscious thought plays no casual role in the execution of any act. It just seems to.

Epiphenomenalism is a hot topic in contemporary philosophy of mind. I will say only that sometimes, in certain cases, the epiphenomenalist suspicion is warranted. The widely accepted view that we sometimes "rationalize" our actions in self-serving ways as we are carrying them out or, more usually, after the fact, is based on the idea that we fabricate rationales that were not really the reasons for our action. Furthermore, as we will see later, mind science has shown that in certain cases we dramatically overweight some reason in explaining our actions, while more important reasons go largely unnoticed.

That said, it is what James called an "unwarrantable impertinence" to think that our conscious thoughts normally play no important causal role in what we choose to do.

Finally, both Aristotle and Dewey argued that at a certain point in development a person has character or personality, a set of relatively stable dispositions or habits of thought, feeling, perception, and action, including various virtues and vices, aims, goals, and projects. Character or personality is conceived as resulting from the complex interaction of an individual with certain innate dispositions with a natural and social environment, and that results in the creature that is me. There is no need to assume, indeed there is every reason not to believe, that who or what I am is ever permanently fixed. I have a personality or character, but who I am, what I am like, changes over time. I am a being-in-time. But there is every reason to think that when I act voluntarily, I do what I want, and that I thereby express myself. Sometimes we act out of character, and normally we regret this. But normally we act in character.

If what we want from free will includes self-control, self-expression, individuality, rational accountability, and political freedom, then

Aristotle and Dewey's naturalistic picture of voluntary action can deliver those goods, that is, five scorecard items. In the case of political freedom, both thinkers see that good as involving our ability to act as we wish in a free society. Whether we live in a free society depends not at all on whether we are metaphysically free. Nor does political freedom depend simply on our possession of the natural capacities to think and act voluntarily. While it is a necessary condition of political freedom that one possess such capacities, genuine political freedom depends largely on the degree to which the social environment allows a person to do what she wishes to do.

Can the naturalistic picture on offer from Aristotle and Dewey deliver the rest of what we want: unpredictability, the ability to do otherwise, and moral accountability? Yes, but this remains to be shown.

Causation

I've said that contemporary mind science assumes causal determinism as a regulative ideal. But it might be better to say that mind science assumes the universality of causation, but is neutral on the question of whether all causation is deterministic or whether some causes operate deterministically and some indeterministically.

In point of fact, when I ask my mind scientist friends whether they assume determinism when doing mind science, they usually answer "yes"—just as Freud and James would have predicted. But when I ask them whether they mean to rule out genuinely indeterministic causation, inherently stochastic processes, most will quickly admit that they don't mean to do so, especially given that statistical generalizations are pretty much all they have. The question arises: Why do mind scientists regulate inquiry by an assumption that is stronger than required and for which, in addition, they cannot produce any real evidence? Remember that we have no laws of mind that can be stated in terms of necessary and sufficient conditions. The answer, as I said in the last chapter, is that by assuming the strongest type of causation, one is motivated to keep on looking for as much causal information as is there. Given some complex outcome, the search for necessary and sufficient causes will yield knowledge of whatever cause and effect relations obtain—even if, at the end of the inquiry, these are only statistical.

There are two senses of indeterministic causation that need to be distinguished. One is *epistemic*. It has to do with our capacities to gather relevant information. One might say that the outcome of the election of the mayor of Hurdenberg is not known until the louse leaps, and where it will leap is an utterly random matter or, what is different, inherently probabilistic. In the epistemic view, the specific conditions that determine the louse's behavior are believed to exist, only we do not know them. If there were an omniscient scientist who knew all the laws of nature as well as the complete state and starting position of the louse, the pheromones emanating from each beard, and so on, she would know exactly which beard the louse would choose.

The other sense of indeterministic causation is *ontological*. According to the ontological interpretation, we would say that the entire state of the universe, including the entire state of the louse, even if known, would not tell an omniscient observer which beard will be the louse's home. We might be able to specify certain objective probabilities *q, r, s, t, u, v, w* that denote each beard's chances of gaining the winning vote. But that's all we can possibly do. And the reason is this: There are certain event transitions, in this case the transition from being a louse-on-the table to being a leaping-louse, that do not have a determinate set of causes that are sufficient to make them go in one and only one direction.

Contemporary quantum physics employs objective ontological indeterminacy. It assigns certain objective probabilities to orbital shifts of subatomic particles that, unless there are as yet unknown "hidden variables" (which, at least according to the favored Copenhagen interpretation, there cannot be), are not deterministically caused. Many physicists say that quantum indeterminacies cancel out above the quantum level so that larger objects behave, for all intents and purposes, deterministically. But suppose this were not true. If it is not true, then quantum indeterminacy, as my colleague Robert Brandon puts it, "percolates up." And thus for some neuron's firing, or even for the activation of two competing sets of cell assemblies, there is no certain outcome. Ontological indeterminism is the thesis that the causal history of the universe includes inherently random, utterly unpredictable event transitions, not because we lack certain information but because there exists no set of causes *in nature* sufficient to determine one outcome rather than another—although, again, we may be able to assign firm probabilistic values to each outcome.

Now I don't claim to know whether, when it comes to persons, the right kind of causation to assume is deterministic or indeterministic. Nor do I think anyone else knows. There is work nowadays in chaos and complexity theories and in self-organizing dynamical systems theory that suggests that the human nervous system operates, at least sometimes, in ontologically indeterministic ways.

Suppose a high school student has been accepted to Duke, the Harvard of the South, and Harvard, the Harvard of the North, and that she is having real trouble deciding which to choose. She can't seem to break the mental tie. Suppose we survey the state of her brain as she deliberates and we see two cell assemblies, one fighting for the Harvard of the South and one for the Harvard of the North, that are of exactly equal strength. We know she must eventually choose. If you believe in strict causal determinism you think something will eventually happen that will tip the balance, and whatever that is will itself have a set of sufficient causes that made it happen.

Suppose the morning sun shines through the window, which causes her to have the thought that she loves sunny weather, and she chooses the Harvard of the South. Or suppose the newspaper on the coffee table mentions that George W. Bush went to Harvard Business School. This impresses her, and she chooses the Harvard of the North. Under each scenario, the shift in her thinking is rooted in consideration of a new reason favoring one option over the other.

Another possibility is that there are states of certain neurons in the relevant cell assemblies in her brain, or of other ones in her brain, which can interact with the ones vying for Duke and Harvard, that are in statistically indeterminate states (there are no laws of nature that can, in principle, predict with perfect precision the next firing state of these assemblies or of certain individual neurons in them). But some neuron or some set of neurons fires, the tie is broken, and she confidently signs the letter accepting the Harvard of the South.

No one knows which picture is truer. Maybe sometimes one obtains, sometimes the other. When push comes to shove (the expression is telling in a number of ways), most people will prefer the first story because it seems to involve completely rational deliberation, whereas the second involves tie breaking by some utterly random process. But it may be a mistake to think that the first story involves significantly more rationality than the second. After all, the fact that the sun came through the window depended on certain utterly arbitrary facts such as that there was a window, that she was sitting next to it, that it was sunny, and so on. Furthermore, the fact that the sun's warmth makes her think she prefers warm weather doesn't make it true. It seems so at this moment, but perhaps it is not, in fact, the case. Perhaps she is not the kind of person who cares much one way or the other about the weather. But even if she does, a wise counselor might say, "Be careful not to overweight such things as the weather when choosing between colleges." Although she has broken the tie by adding a new reason to the mix of reasons she was already considering, we might not judge her final decision to be all that rational.

A strict causal determinist might say that although this individual deliberated and made her choice, it was inevitable that there would be a window, that the sun would shine when it did, that warm weather thoughts would move her away from the deliberative tie, and also that it was inevitable that there would be no wise counselor available to advise against overrating weather. No one will doubt that she made a reasonable choice—she had two excellent options, after all. But again, we might rightly shy away from saying that her choice, indeed any human choice, was or is completely rational.

Things aren't all that different if one is a causal indeterminist. We begin with the very same picture of the individual weighting the pros and cons of each college, and then some random neural discharge tips the balance. Again, the tie is broken, and it is not clear that there is less rationality in this case than in the first. To be sure, in the first case she has one more thought than in the second, namely the one about the weather. But what caused that? According to the determinist, the sun's shining was sufficient to activate the thought that she prefers warm climates to cold ones. But, again, this thought may not have led to the most rational decision. In the second case, no weather thought occurs, but she finds herself, as we all often do, feeling the weight shift. Duke it is. Both cases involve rational deliberation. But it is not obvious that the first kind of deliberation involved greater rationality than the second.

Thus the ideal of complete rationality may be as much a fiction as the idea of a completely unconstrained will. The fact that we are rational animals does not remotely entail that the only things that move us are "good reasons" or, in many cases, things worth calling "reasons" at all.

Dialectical Space

The traditional philosophical conception of free will involves deliberation, reasoning, and choice that are not causally constrained either deterministically or indeterministically. The scientific image assumes causation. Quantum physics, and some higher-level theories of the behavior of self-organizing systems, assume causal indeterminacy. Certain things happen without there being any set of sufficient causes that make them happen. That said, there is considerable disagreement as to whether the indeterminism insisted upon by these latter sciences is epistemic, a problem with our knowledge due to nature's complexity, or genuinely ontological.

Since science has yet to resolve to its own satisfaction where to divide the world between those parts governed by deterministic causal processes and those governed by indeterministic causal processes, I will sensibly plead ignorance about where that divide plants itself. Some sciences, quantum physics for example, do best assuming causal indeterminacy; others, such as astronomy, do best by assuming causal determinism.

Furthermore, unlike many of my colleagues I don't think any important conclusion about the Cartesian conception of free will turns on resolving this matter. If determinism is true of persons, then for each act of will there exists some set of prior or contemporaneous causes that are sufficient to cause the will to be in the state it is in, that is, to will as it does. My deliberation, reasoning, and choice are constrained by these causes, whatever they are. If, on the other hand, causal indeterminism is true of persons, or more likely of certain goings-on inside the central nervous system of persons, then my will is in the state it is in because certain prior or contemporaneous states of affairs, with certain objective probabilities of occurring, did occur, and so shifted my deliberative, reasoning, and choosing processes from one state to another.

As long as causation is ubiquitous, whether deterministic or indeterministic, my will is never in a state that is not affected by prior causes. There can be no such thing as an act of will that is the act of a "prime mover unmoved," or a will that is "so free, that it is completely unconstrained."

The idea of a completely unconstrained will is thus inconsistent with the scientific image. But I want to claim something stronger still. The idea of agent causation is incoherent independently of what the scientific image says.

When a philosophy student learns about the problem of free will, she is introduced to the problem of "free will and determinism." But if what I have said so far is right, it would be better if she were introduced to the problem of "free will and causation," or even better, to the problem of "deliberation and choice and causation" or simply the problem of "the voluntary and the involuntary." The reason that framing the problem in terms of free will and determinism is a bad idea is that neither concept adequately points to something that is credibly at stake. The problem, assuming there is one, is with the alleged ubiquity of causation, be it deterministic or indeterministic. The only conception of free will that has ever been entertained that deserves the name of free will is the Cartesian conception of a mode of mental processing, or a mental faculty, that is totally unconstrained, totally self-caused. The prime mover, itself unmoved. But there is no reason, none, to think that there could be any such thing. It is so conceptually puffed up that it is incredible, incoherent.

Consider what it would mean to have such a free will. When I make a choice I do so *ex nihilo,* by electing, without anything constraining my deliberation, a course of action. But if nothing constrains my choice, then reasons don't constrain my choice either. And if that is so, then ordinary introspection must be deemed wildly wrong. After all, it seems to most everyone that when they are deliberating among the options at hand that they are weighing pros and cons and that this information constrains the choice.

Second, and just as bad, if when I choose I do so for no reason (choice may create a reason for action but does not itself rest on any reasons) then my choice is either arational or irrational.[5] Since one of the main things—perhaps *the* main thing—any conception of free will worth wanting is supposed to do is to explain how rational choice is possible, and so to explain how I can be held rationally accountable for my choices, the orthodox conception of free will is a miserable failure. It is conceptually incoherent, in the sense that it provides no coherent way of conceiving of what it wants to gain for itself.

If you were only able to say that the orthodox picture of free will makes no sense from the perspective of the scientific image, you could be rightly accused of begging the question. All you would then be saying would be that what I assume doesn't permit what you assume. But I am making a stronger claim. Upon examination, the orthodox concept of free will makes no sense in terms of the agenda it sets for itself—to explain rational deliberation and choice.

If this is true, then there is no problem of "free will and determinism" worth discussing. There is a problem in the vicinity worth discussing, but free will and determinism is not it. The problem worth discussing is how to make sense of freedom, deliberation, reason, and choice within the framework set out by the human sciences generally and by mind science in particular. This can be done.

But before proceeding, let me mark as clearly as I can where I am positioning myself. The texts that discuss the "free will–determinism" problem take two main positions:

- *Compatibilism*: Free will is compatible with causal determinism. Most compatibilists say that free will requires causal

[5]There might seem to be a way out. But it is not worth taking, at least not by an advocate of the view of free will under discussion. The way out I am thinking of would involve saying that when I choose some course of action now, I do in fact consult my self-chosen past reasons. The trouble with this tactic, in addition to conceding that for any actual deliberative episode my past reasons constrain it, is that in order to make the move *and* to preserve the unmoved mover picture as the view it is, the proponent of that view must think that the first choice or reason in the relevant deliberative chain was arational or irrational by virtue of being uncaused or caused by no reason at all.

determinism in the sense that the state of my will (itself determined by prior and contemporaneous causes) must be a sufficient cause of any choice I make.

- *Incompatibilism*: Free will is incompatible with causal determinism. Incompatibilists take one of two roads. *Libertarians* claim that since we have free will, determinism is false. Libertarians employ the concept of free will as Cartesian agent causation, or those who sense its incoherence by a promissory hand wave in the direction of "something or other that does the trick but that is yet to be articulated or formulated to anyone's satisfaction." *Hard determinists* claim that since determinism is true, there is no such thing as free will.

Given my argument that the normal way of framing the problem—as the problem of "free will and determinism"—makes no sense, you will not be surprised to discover that I think all these answers are unsatisfactory and the reason is that the problem is ill-posed. If forced to comment on the three positions I would say this. Libertarianism is a nonstarter because the Cartesian conception of free will, the only conception that has received articulation within philosophy as deserving the name free will, is a nonstarter.

The compatibilist, meanwhile, if he thinks free will is compatible with determinism, must have changed the subject. He cannot be saying that the Cartesian conception of free will is compatible with determinism because, well, it isn't. And indeed if one looks at the literature one will see that compatibilists invariably mean something different by free will than what the orthodox concept says it is.

The hard determinist, unlike the compatibilist, accepts the terms of the exercise as they are set and sees correctly that determinism is incompatible with free will, as the Cartesian conceives it. But both the compatibilist and the hard determinist make the same mistake. They both claim to *know* that determinism is true. But if what I have said about causation—there being both deterministic and indeterministic causes—is plausible, then neither can sensibly be said to know that determinism is true. Causation is ubiquitous. Ours is a causal universe. But no one yet knows the exact range of deterministic and indeterministic causation—assuming the universe contains some of each.

What to do? My proposal is this: Change the subject. Stop talking about free will and determinism and talk instead about whether and how we can make sense of the concepts of "deliberation," "choice," "reasoning," "agency," and "accountability" (scorecard items) within the space allowed by the scientific image of minds. This is, I hasten to admit, just what I accused the compatibilists of doing. Since they cannot be saying that free will is compatible with causation, either deterministic or indeterministic, they must be claiming that something else—hopefully something similar to free will—is compatible with causation.

It would be misleading to call my position compatibilism, however, since compatibilism seems to accept the terms of the standard debate about "free will and determinism." Since I have been trying to frame the pressing question in terms of the compatibility of "rational deliberation and choice and causation," or as the problem of the voluntary and the involuntary, it will be best to call my view *neo-compatibilism*.[6] I do claim that we can make sense of rational deliberation and choice in a causal universe.

Statistical Generalizations: A Cautionary Note

Having noted that there may be both deterministic and indeterministic causes, I must take a couple of pages to post two cautions. First, some have thought that casual indeterminism might be enough to save the Cartesian conception of "free will." This is not true. Second, many people take comfort in statistical generalization because they think that no absolutely certain predictions follow from such generalizations, which is true, but also because they think that such general-

[6]I don't intend by calling my view neo-compatibilism to suggest that my view is really new. It is mainly a way of marking and acknowledging the fact that my view, just like the views that inspire me—those of Harry Frankfurt, Gary Watson, John Martin Fischer, Susan Wolf, and Daniel Dennett to name a few—does not in fact agree that the traditional concept of free will is compatible with determinism. These fellow philosophers think, truly, that some suitably naturalized conception of human agency preserves some, but not all, of what is worth preserving in the traditional concept of free will, and that this concept is compatible with a causally well-behaved universe.

izations imply causal indeterminism. There may well be causally indeterministic processes, but the fact that many scientific generalizations are statistical doesn't imply that the processes that are explained statistically are not deterministic. Let me take each point in turn.

Quantum physics, as I've said, reveals indeterminacy at the lowest levels of the material world. It isn't just that we cannot reliably predict when an electron, for example, will shift its orbit. Even if we had omniscient knowledge of all conditions affecting the electron, we couldn't. The type of causation that governs subatomic particles is inherently random, or better, stochastic. Electrons shift their orbits without being caused to, even though we can precisely specify the relative probabilities of orbital shifts.

You will not be surprised to hear that a small philosophical industry was launched and flourished, from the 1930s through the 1960s, as much as a misguided fledgling enterprise could be expected to do, which tried to explain human capacities of free agency along quantum mechanical lines. More recently, in the last two decades, the emerging sciences of chaos and complexity and self-organizing dynamical systems have been applied to the brain. And it has been suggested that the brain, like the weather, is inherently unpredictable (which is almost certainly true), that the brain like the weather is a self-organizing dynamical system (also almost certainly true), that the brain like the weather is exceedingly sensitive to extremely small perturbations (true again), and that certain brain processes are inherently random (maybe).

As I have said, I am open to there being genuine ontological indeterminacy at both the quantum level and the level of neural processing. But the attempt to gain free will from indeterminacy at the quantum level or at the level of global brain processes is a bad idea. The last thing anyone wants is for free will to be the result of random causal processes. Give me the choice between my actions being strict causal outcomes of my genes, my history, my personality, the state of my mind/brain, and the current environment, and I will take that, every day of the week, over the view that when I deliberate or act, I do so randomly, because my will has flown the causal coop and moved arbitrarily into a new, unpredictable place in the causal nexus.

In 1934, Hans Zinsser published his classic *Rats, Lice, and History: A Chronicle of Pestilence and Plagues,* from which the story about mayoral elections in Hurdenberg is taken. By telling his history from the point of view of bugs, germs, and vermin, Zinsser effectively destroyed the notion that history could be told solely, or even primarily, in terms of the acts of great men. He calmly writes, "It is now quite well established that the subject of our biography [human history] is, in some phases of its adventurous existence, closely associated with rats." Elsewhere he rightly complains that the *Encyclopedia Britannica* devotes half as much space to the louse ("this not unattractive insect") as to the nearby entry for "Louth, a maritime county in the province of Leinster," and one-fifth as much as it devotes to Louisville, Kentucky.

Not only does the encyclopedia not mention the louse's political importance to Hurdenberg, Sweden, but "this creature which has carried the pestilence [typhus] that has devastated cities, driven populations into exile, turned conquering armies into panic stricken rabbles, is briefly dismissed as a 'wing-less insect,' parasitic upon birds and mammals, and belonging, strictly speaking, to the order of Anoplura."

There is no doubt that human history is extremely sensitive to small perturbations and random events (no matter whether the randomness is epistemic or ontological). But we don't want our free will to work that way. Just as I don't want my will to be governed by the random forces that cause a subatomic particle to switch its orbit, I don't want some act I elect to perform to be the result of the neural equivalent of the utterly arbitrary act of the louse who elects the mayor. This, I think, effectively takes care of the strategy that sees hope for saving the traditional notion of free will by conceiving its operation along indeterministic (epistemic or ontological) lines.

A superficially more promising strategy is to point out that most laws in mind science, indeed in the human and biological sciences generally, are statistical or probabilistic, and that such laws leave the traditional conception of free will space to operate. Laws in population genetics tell us that if there is a mutation of such and such sort, and the mutation subserves a phenotype that leads to differential reproductive success for the plants or animals that possess it, then the trait will increase in frequency in the population. The law of operant conditioning in psychology says that if a particular behavior is rewarded then that behavior will increase in frequency. Neuroscientists talk about the normal development and function of the visual or auditory system and about the increased chances of getting Alzheimer's disease if you have the APOE-4 gene. Clinical neurologists and neurosurgeons never attempt to spell out the complete causal chain leading to visual or auditory dysfunction or to a particular case of Alzheimer's.

The lesson sometimes taken from this is that statistical or probabilistic regularities reveal patterns and influences. And patterns and influences might be compatible with the Cartesian picture of free will (although I think even this is a potentially destabilizing concession): When I choose, I do so freely, albeit subject to certain influences.

The trouble with this thinking is subtler than the trouble with the attempt to gain free will from quantum indeterminacy or chaotic unpredictability. It is in the nature of laws governing populations, populations of genes, or a population of responses available to a particular organism that they are statistical. The events leading to a particular mutation or behavior, however, are typically thought to be governed by strict causal processes, even if no one cares or can discover what exactly those events were. In effect, when we say that rewarding a person for truth telling will increase the probability of that person's telling the truth in the future, the psychologist doesn't normally think the exact increase in probability is subject to random forces. She is assuming that you have a repertoire of behaviors, all with different probabilities of being displayed due to their prior reinforcement histories, and that any particular behavior displayed or enacted is completely determined. She speaks in terms of probabilities only because, since the determination involves your entire life history, the exact reason for that particular response may well be unknowable.

Although it is not absolutely mandatory, most advocates of the scientific image assume that all thought and action is determined. Problems in specifying the exact causal determinants in a particular case, or predicting precisely what will happen, are typically believed to be problems of knowledge, not problems with nature.

When a doctor tells a patient that he has a certain type of cancer and that he has a 50 percent chance of survival, the doctor is telling the patient that among a population of patients with cancers of that type, 50 percent survive. But does, or should, either the doctor or the patient think that which outcome occurs will not in the end be strictly determined? No. At least, not necessarily. The patient falls into a 50/50 group given his type of cancer. There is lots of missing information—information about spread, about the particular patient's immune system, and so on. Which outcome occurs will be causally determined based on the surgeon's ability to get the whole tumor, and on all of these other factors. What doctors and their patients call "miracles" are just rare causal outcomes, they are not uncaused.

The main point is that statistical generalizations do not imply that the causal processes in question are genuinely indeterministic. Some might be, if causal indeterminism is true in those cases. But in certain cases, such as those just discussed, the statistical nature of the generalizations can be explained in terms of epistemic limitations.

Often the statistical nature of a generalization is not due to any ignorance—indeed we might know everything relevant—but simply to the fact the generalizations are about the behavior of a whole population. In such cases, the explanation starts by specifying the degree to which different members of the population display some set of characteristics with 1 being the highest possible score for any individual member assuming she possesses all the relevant traits. Suppose we seek to explain and predict the future distribution of red fur and blue eyes among some population of creatures that is now comprised of 50 percent with both traits and 25 percent with red fur and brown eyes and 25 percent with black fur and blue eyes. Even if there are known deterministic laws governing fur and eye color and even if we completely control interbreeding, it is in the nature of how traits are redistributed in populations over time that all we can say is that for generation g+1 the population will be at 60–40, and for g+2 it will be at 70–30, and so on. It is not as if we are only 60 or 70 percent confident in our prediction. We are 100 percent confident in it. It is just a (deterministic) fact about this population that it will move first to a 60–40 distribution and later to a 70–30 one.

The first tactic in trying to make room for the traditional conception of free will in terms of causal indeterminism fails because it makes free will subject to nonrational random forces. The second tactic is somewhat different. It takes whatever openness statistical generalizations offer and proposes that somehow, within that open space, free will performs its magic. There are two problems with this idea. First, the fact that a generalization is statistical does not entail that the underlying process is not deterministic (although it might be). Second, no one knows how to make sense of the idea of an act of will that is both not determined and, at the same time, rational. We do understand what it means for a causal transition to occur indeterministically. But indeterministic transitions are not rational. No cause, including no rational cause, forces whatever event transition occurs at the subatomic level to make a neutron decay into a proton and an electron.

We also understand what it would mean for a person to deliberate or act rationally. Doing so requires that she consider the reasons she has for thinking or acting in a certain way, and if she reaches some conclusion in thought (say she is trying to solve an arithmetic problem), or decides to act (she decides to go to the concert instead of the movie), she does so on the basis of her reasons. But nothing in this account of rational deliberation and rational choice leaves space for a completely unconstrained act of will, for a prime mover unmoved. To be sure, I am moved by my reasons. But the reasons available to me are a product of what the world, and my history, allows me to see and consider, given my cognitive equipment and capacities. The problem I am trying to solve or the decision I am trying to make isn't one I choose or create *ex nihilo*. I am trying to solve a problem or make a decision that has arisen because of who I am, what I am like, what I am now doing or thinking about, and what the social or physical environment has sent my way as options. It is not clear where free will comes in, or if it did, what it would be doing.

Explanation, Prediction, and Indeterminacy

One reason people don't like messing with free will is that they think that the claim that causation is ubiquitous, especially if it is deterministic causation, means that the future course of their lives can be predicted. It is disturbing to think that my life story could be told before I live it out.[7] But this eventuality should not worry us. Whether causal determinism is true or not, complete predictability is not in the cards.

"El Niño" involves the Pacific warming by a just a little—a couple of degrees Fahrenheit—in the vicinity of Australia and Indonesia. If the trade winds blow a bit farther southward than normal, then the warm ocean water spreads to the shores of Peru, coastal productivity plummets and fishing stops, terrible floods occur in South and Central America, drought hits the North American Pacific Coast, tropical storms are stirred up in the Gulf of Mexico, and the South Atlantic produces fewer but more furious hurricanes than usual that hit the East Coast of the United States. There is more, but you get the drift.

My suspicion is that were I to ask a trained meteorologist about El Niño, she would explain to me that the weather is thought to be subject to "random, chaotic forces." El Niño itself being an example. And I suspect she would think that the relevant forces made the weather, in principle, unpredictable. One might understand the distinction between in practice and in principle this way. There are countless reasons why the water near Australia and Indonesia might get warmer. An underwater volcano might erupt. Lighting might strike a drought-ridden area of the Australian outback, millions of acres might burn, diverting wind currents that would normally cool the water, and so on. In theory, meteorologists could track volcanic eruptions and wildfires (perhaps they already do)—but perhaps the relevant water warming could be caused by a malfunction of the sewage system on an ocean-liner that caused its effluent to be dumped in the wrong place at the wrong time. Again in theory all ships could be tracked. But the idea of tracking all possible causes is practically crazy.

Suppose, going back to volcanic eruptions or to the lightning strike in the outback, that our best science says that these things are inherently stochastic. Probabilities can be assigned to subterranean lava flows within certain ranges and to lighting strikes hitting a cer-

[7]Although almost no one seems bothered by the fact that divine omniscience might mean that God knows how my life will go. The troubling worry is that my life story might be revealed to me before I live it, or that other people—the mind scientists of the future—will be able to say how it will go.

tain tree. But even an omniscient scientist could not predict the relevant events with precision. They are, in principle, unpredictable since these phenomena only abide stochastic laws, not strict deterministic ones.

This seems perfectly credible, and it might be thought to bode badly for the mind scientist who regulates her scientific inquiry with the assumption of strict causal determinism if certain mental processes are inherently indeterministic in the way some think the weather is. Even as a regulative ideal, there may be a problem with making the assumption that both Freud and James thought the psychologist must make. If there are inherently stochastic processes at the level she studies, then perhaps she should make a weaker assumption than that of strict causal determinism.

Maybe, maybe not. I've been insisting that it's not unreasonable in a situation of vast ignorance to assume strong causal determinism to see what firm generalizations the human mind might be made to yield. If there comes a time when theory and evidence make it plain that we are dealing with some processes that are inherently indeterministic, then, following James's insight that the assumption of determinism is provisional, methodological, and revisable, it will be wise for mind scientists to weaken their regulative assumption for whatever phenomena some weaker, namely indeterministic assumption reveals itself to be applicable to. But it is too early to say that this should be done now.

It is important to keep the issue of prediction and explanation separate. Explanations occur after the fact, whereas prediction occurs in advance. Once an El Niño arises, even if unexpectedly, we can typically explain how it got started and predict, but only within a range, its expected effects. Likewise, if a lightning strike causes a wildfire that destroys a house, we can easily trace the destruction back to the lightning strike and the preceding drought. But even if we could give the statistical chances of the house's burning before the fact, we could not have predicted the particular lightning bolt, the particular day, or the particular wind gusts that drove the fire toward that particular house. Once we admit the possible indeterminacy of inherently random events—such as lightning striking, typhus epidemics caused by lice, or random neural firings—that cannot in principle be predicted in advance, the specter of complete predictability vanishes.

The difference between explanation and prediction is not only important philosophically, but it can, in my experience, bring some consolation to those who think that if we can explain human thought and action, we will be able to predict it completely. Trust me, this is not in the offing, now or ever. So unpredictability, another scorecard item, is satisfied.

If there is such a thing as free agency or voluntary action, it cannot, if the scientific image is true, be immune from the causal laws that govern all things physical. Or to put it more moderately, even if strict causal determinism is too strong an assumption and there are genuinely indeterministic—random or inherently stochastic—processes in nature, these give no solace to the libertarian. Random, chaotic or stochastic processes are not what the libertarian is seeking. He seeks a theory of human action according to which the agent makes choices that are self-caused. Free actions, if there are any, are not deterministically caused nor are they caused by random processes of the sort countenanced by quantum physicists or complexity theorists. Free actions need to be caused by me, in a nondetermined and nonrandom manner.

This is why the two components of the Cartesian picture are natural partners. Natural phenomena obey deterministic (or possibly indeterministic) laws. Conceive the mind as nonphysical and it need not obey natural laws or operate according to natural principles.

This, of course, is a terrible idea from the point of view of a defender of the scientific image. She countenances no such realm as that of the nonnatural. Is there any space to move, any space left within the scientific image for the idea of voluntary action, for a naturalistic notion of free agency? The answer, I've said, is "yes." Whether the naturalistic picture I advocate requires the defender of the manifest image to concede more than he is willing is, in my experience, a variable matter. When it is, he is asking for more than he can have. It is not simply that you can't always get what you want. He wants something he shouldn't want, something both unnecessary and incoherent.

Information and Causation

Behind Descartes's ill-conceived idea of free will is the notion that mental causation is fundamentally different in kind from ordinary causation. And indeed it is. To speak in the language of cognitive science, reasoning involves computation, or what Thomas Hobbes called "reckoning," over mental representations. In fact Hobbes said that "by ratiocination I mean computation." He did not mean that thinking consists in doing arithmetic, but that thinking involves the manipulation of what he called thought "parcels" and "phantasms" in accordance with rules. However exactly reasoning, deliberation, and choice work, they involve the manipulation of thoughts with meaningful semantic content. Thought, deliberation, and choice operate by way of information processing, some portion of which involves consciousness.

The idea of a self-caused will was articulated in contrast to the world described by Galilean and Newtonian mechanics, in which inanimate objects move because they are thrown or propelled by other inanimate bodies. Galileo was an expert on catapults. And the world described by mechanics is a world in which, paradigmatically, the concept of cause was pictured in terms of objects of considerable size forcing other objects to move. It is no accident that when Descartes tried to explain bodily processes such as muscle contractions, he posited small distilled particles of blood that literally fill the muscles. But one might sensibly wonder why this inventor of analytic geometry didn't see more promise in a conception of deliberation and reasoning as information processing, since he surely recognized that doing mathematics involved symbol manipulation in accordance with strict transformation rules.

Perhaps Descartes did not see the promise in this idea because the standard concept of cause was so conceptually tied up with the idea of collision. In the seventeenth century, when the idea that human thought and action might be part of the natural fabric of things and thus subject to natural law began to be openly considered, the dominant idea that causation is collisionlike made mental causation almost impossible to envision. One might have considered the idea that reasoning, deliberation, and choice operate in accordance with collisionlike causal processes, as Descartes almost certainly did. But it sure doesn't seem as if they do. And thus the idea of mental causation as information processing failed to gain a foothold within the conceptual space of that which is caused.

The modern sciences of information theory, computer science, artificial intelligence, and cognitive science have made clear that certain causal processes are best described in terms of information flow, not in terms of collisions. And in this way they have shown—with considerable help from the sciences of magnetism, electrodynamics, particle physics, and chemistry—how to eliminate the idea that all causation involves the collisions of bodies, or even particles.

A computer program operates by symbol crunching. A program is a set of rules that governs the behavior of a system capable of detecting what symbols (normally electrically encoded or represented) are presented to it. The system manipulates these symbols in accordance with the rules embedded in its program, performs whatever computation is called for, and moves to the next state to perform its next task, as determined again by what it detects and by what rule applies.

You can read about the information sciences in hundreds of other books. All I want to say for present purposes is that the existence of these sciences, and the actual existence of computers, is proof that there are systems that do their thing by manipulating symbols in accordance with rules. A guiding idea of contemporary mind science is that we are such a system. We are systems that function by, among other things, manipulating and processing information.

Now one very big difference between humans and computers, as we now know them and can build them, is that we are consciously responsive to information. No one has a clue as to how to build a conscious computer although we may, in fact, be one. That said, there are many different computational architectures. Some computers operate serially according to strict "if-then" algorithms, of the sort that produce perfect solutions to mathematical problems given certain axioms. Since not all problems humans need to solve are mathematical, such models are almost certainly not suitable to model real minds in all domains. Computers programmed to solve real-world problems are typically equipped with heuristic rules that help sort out what information is relevant and what isn't, given some problem or other.

A computer designed to help solve all problems associated with a certain satellite, when faced with an electrical irregularity six months after its launch, needs to know not to consider the ground temperature at the time of launch as a plausible contributing factor.

The standard view nowadays is that if the mind is a computer, it is one composed of many minicomputers in massive interaction. The mind, most everyone thinks, is a massively parallel system. Marvin Minsky's picture of the "society of mind" was one of the first proposals that we think of mentality as constituted by the interaction of numerous, possibly many thousands of little processors each doing its own thing and passing the information it computes to wherever it is needed.

Connectionist models are currently the rage among mind modelers. Connectionist systems comprise numerous processors, each with an initial probability of being activated in one way rather than another. Such systems have their weights reset by receiving input, giving output, and then receiving feedback about whether they are moving in the right direction as measured against some external standard. NETTALK is a justly famous connectionist system that starts out knowing nothing about the pronunciation of American English, although it has the ability to produce all manner of speech sounds. As it hears American English, tries to pronounce words for itself, and receives feedback about how it is doing, the initial weights shift. The initial weights can be understood as its initial preferences regarding the noises it is disposed to make, whereas subsequent weights can be thought to model development. NETTALK is made of plastic, silicon, and circuitry and yet it learns. Its initial settings embody an incredible degree of freedom, for depending on what language it hears, it—or some more sophisticated descendant—can, in principle, learn to pronounce that language. In any case, after many, many hours of training, NETTALK speaks in a way that we can recognize as American English.

It may be that to model the mind computationally, we will need to weld together all the known computational architectures, as well as others not yet invented. Still, we now understand how causation can operate without collisions of the Galilean–Newtonian sort. It can do so via information processing. The electrical transitions involved in computation information processing are causal, but not remotely collisionlike.

Two final points. First, I have said that no one knows how to build a conscious computer, indeed no one is even trying. This thought might give comfort to the epiphenomenalist who claims that consciousness does no interesting work. But the fact that computers can do their thing without consciousness has no implications for the role consciousness plays in human mind/brains. Humans are conscious, and the most plausible starting assumption is that we evolved to process information consciously for good biological reasons. The burden of proof falls completely on the epiphenomenalist to show that consciousness plays no important role. Second, and relatedly, I claimed that one of the things we want from any conception of freedom of deliberation, reasoning, and choice is "reasons-sensitivity." We now have that. Reasons, as commonly understood, are just a type of information that serve the conscious mind/brain as causes of action. I have a reason for action if I am hungry and see that there is an apple in front of me.

In any case, it is a standard assumption in both contemporary philosophy of mind and cognitive neuroscience that reasons can be causes. If this is right, then even if computer science at best provides analogies for mind, we have an understanding of how reasons-sensitivity is possible. And reasons-sensitivity was on our checklist of requirements for an acceptable conception of free agency. Rational deliberation, the next requirement on our list, is best conceived as the process of building an overall rationale for some conclusion or course of action by blending together the relevant information in a principled manner, so as to yield a sensible conclusion or choice. Again, if we conceive of causes not as collisionlike events but as algorithms or heuristics, sets of rules for dealing with information, we can see how rational deliberation is possible for complex creatures, artificial or natural.

There are only two requirements left on our initial list: the capacity to do otherwise and moral accountability. Before I show how these are to be won, I should say a bit more about self-expression, deliberation, reasoning, and choice as they are involved in creative thought, and then with that behind us, to speak specifically about how moral agency and the capacity to do otherwise work.

Creative Agency

Whether or not one buys into the received view of the nature of free will, most everyone thinks—and thinks rightly—that questions about free agency matter. They matter to the way we make sense of our own actions and those of others. They matter to our practices of reward and punishment, approval and disapproval, and to the meaning and significance of our lives. It is important that I, to some significant extent, am in charge of my life, that my choices are self-expressive, and that they are self-expressive by virtue of being my choices.

In the last chapter I warned of a certain kind of fear-mongering, as exemplified in Nagel's worry that if I am just one part of the causal nexus, then my agency shrinks to an "extensionless point." I pointed out that this is hyperbolic. Everything involved in the causal nexus is responsible for producing certain effects. Nothing causal, including human agency, is reduced, no matter what perspective one takes, to an "extensionless point." Skinner plays the same sort of trick in his lecture "On Having a Poem," which describes writing poetry or prose or a lecture in a way that dramatically underestimates an agent's own contribution in creative production. Skinner says that if his genes, history, and the state of his brain had been different, then his lecture would have been different. Probably true. But what is misleading is, first, that Skinner thinks this warrants the analogy of writing a poem, book, or lecture to having a baby and, second, that the words "the state of my brain" say enough to stand for, indeed to explain, his entire agentic contribution. The problem with the analogy of writing a poem or giving a lecture with having a baby lies in the word "having." To be sure, becoming pregnant and having a baby occasion many, many thoughts. But these thoughts play no interesting causal role in the processes of gestation and birth. We may have babies, but we produce poems and lectures. Second, even assuming mental activity is realized in the brain, no *one* state of any brain (even throwing in genes and history) has ever produced a poem. Writing of every kind requires extraordinary thought, the "to-and-fro" of trial and error, test and retest, excision, revision, and expansion. One's sense of what works, what is clear and unclear, memorable associations, and much else come into play. Furthermore, writing involves as much intuition as preplanning, so we will need to make room for impulses and thoughts that bubble up for reasons we neither consciously create nor completely understand. This book has taken me many hundreds of hours to write. It is preposterous to think that even one sentence in it can be explained in terms of a single proximate brain state.

Behaviorism was notorious, in its heyday, for having nothing to say about mental causation. In part this was due to an epistemological conservatism that required silence when speaking about unobservable processes; still, whatever the reason, behaviorism focused exclusively on inputs and outputs, stimuli and responses. One might think, therefore, that Skinner's hand wave in the direction of the state of his brain in producing a poem or a lecture was a generous gesture that acknowledged that something at least goes on inside a person engaged in writing. But this would be too kind a reading. The process of creative writing is hardly even plausibly gestured at if the gesture is merely in the direction of some state of the creator's brain. Writing involves very complex mental processing, normally over a substantial period of time. There are literally zillions of mind/brain states.

Nagel's and Skinner's problems are similar. Nagel's depiction of agency as an extensionless point is only plausible, and not just for agency but for any causal factor, if we take the point of view *sub specie aeternitatis*, "the view from nowhere," which Nagel criticizes elsewhere, as the view from which nothing stands out from anything else. If, on the other hand, we do not take the "view from nowhere," but the view from somewhere, we normally come poised with a particular question or set of questions about some particular event or set of events. "Why did Kennedy think committing so many troops to Vietnam was a good idea?" Once we target one thing, rather than everything, as in need of explanation, every causal factor in its vicinity can potentially gain our attention, and the causal particularities involved will come into focus so we can give deep and rich texture to what we see as in need of explanation. Nothing in the vicinity of what-needs-to-be-explained will recede to some extensionless point. Quite the contrary.

In Skinner's case, the hand wave in the direction of the poet's or writer's brain state (plus his genes and history) is at best a lame gesture that announces Skinner's commitment to the explanation of even creative work in naturalistic terms. But absolutely nothing informative has been said about how mind/brain processing works for the general case of creative writing (assuming something general and informative can be said about such cases). Skinner has marked nothing relevant to any particular creative act, in this case his writing the lecture "On Having a Poem," that explains even one feature of it. At least he has explained nothing by saying the state of his brain was relevant. As the kids say, "Duh!"

Both Nagel's and Skinner's ways of putting things arouse anxiety by exacerbating the worry that nothing interesting can be said about human agency if causation is ubiquitous (Nagel), or if mental activity is treated as a black box about which we should remain silent (Skinner). But this thought finds fertile ground only if one takes a very abstract, removed perspective, and comes poised with no specific question about some interesting human act. In point of fact, every contemporary mind scientist assumes that there exist causally rich stories—stories of unbelievable complexity—for how persons by way of their minds do the things they do. There is nothing in the image projected by contemporary mind science that denies that we are agents who deliberate, reason, and create. When we do these things, we deploy extraordinarily complex mental skills, and we express our individual identities (again, a scorecard item).

Moral Agency

Morality is not the only part of life in which agency seems essential, but focussing on it will help to see more clearly why so much turns on the issue of free will—and why causal determinism, or any causation, seems so threatening.

Morals are concerned with how I personally, and we collectively, ought to live. Morals, or ethics (I'll use the terms interchangeably), ask such questions as: What is a good human life? How is it best to be and to live? What rules, principles, or commandments must a good person abide? What virtues must a good person possess? Morality has, we might say, intrapersonal and interpersonal sides. How can I be the best person I can be? How ought I to work on my character to be happy, to flourish, to find serenity, to achieve, as best I can, as much personal perfection as I can? On the other hand, how ought I to feel and act toward others? What are my duties to other sentient beings, not just to humans, not just to my loved ones, but to the poor, the indigent, the suffering on the other side of the globe?

Morality concerns itself only with ways of being and acting that are under voluntary control—with matters in which agency, choice, and options are involved.

One might worry that if the scientific image is right, and in particular if human motives, thought, feeling, and action are part of the natural order, then there is no domain of the voluntary, and thus nothing for morals to do.

It seems as if *I* do things. It would undermine, possibly destroy, the meaning and significance of my life if *I* am not an agent, if who and what I am is in no way the result of choices I make. It matters that I am not just along on some ride that the cosmos, for some absurd reason, is taking. And yet if I am just an animal, if what I think and do is just the emergent result of what the world outside, my body, and its brain jointly produce, then it is hard to see in what sense I am an agent, I am self-productive, and I create or co-create my life's meaning.

These are reasonable worries, but they may not be rational. True, it is hard to see how mind science can make room for the idea of free will, because, well, it can't. There is no other explanation for why standard discussions of "free will and determinism" are so disquieting for college freshmen who are exposed to the problem in Philosophy 101. We professors who raise the issue need, genuinely, to be concerned that in dramatically upping the ante on normal late adolescent angst we may be doing some sort of psychic harm. I remain convinced that it would be more intellectually honest to explain the history of that quandary in a way that shows why the conflict was initially set in terms of Cartesian free will and strict determinism, why now setting it in those terms makes no sense, and then to get on with trying to make sense of voluntary thought and action for animals like ourselves. I also think that this way of addressing the problem would reduce, possibly eliminate, the feeling that there is no way out.

In any case, that it is hard to see how the scientific image can make room for free agency does not mean that it cannot do so—that it cannot preserve the ideas that we deliberate and make choices. It can. Although, as we've seen, what room the defender of the scientific image can make is less room than the room a totally free-ranging Cartesian agent thinks she must possess in order for life to have meaning and for morality to be possible. But given the incoherence of the Cartesian concept of free will, this is no longer a serious problem.

The Cartesian picture of agency, and the manifest image that avows it, cannot be meshed with the scientific image for the simple reason that one cannot mesh an idea that makes no sense with ideas that do make sense. But there is, as I have argued, a way to preserve the idea that we are free agents who have a certain amount of authorial control over our lives, and who do make voluntary choices.

Recall the list of conditions I claimed might be plausibly demanded of any credible account of free agency:

- *Self-control*
- *Self-expression*
- *Individuality*
- *Reasons-sensitivity*
- *Rational deliberation*
- *Rational accountability*
- Moral accountability
- The capacity to do otherwise
- *Unpredictability*
- *Political freedom*

The italics denote the conditions I claim to have met or, better, the conditions that I have sketched ways to account for in naturalistic terms. There are still two conditions left in regular type. I now turn to these.

Many believe that it is really and mainly the conditions necessary for moral agency that require the Cartesian picture of free will. If my will is completely unconstrained, then I am morally accountable for what I do, and given that I operate in a completely unconstrained way, I can always do otherwise than I in fact do or did.

There is no Santa Claus, no Easter bunny, and no tooth fairy. But kids, when they learn these truths, seem to adjust. They still get presents at Christmas, painted eggs and candy at Easter, and money for a lost tooth. There is likewise no such thing as Cartesian free will. But maybe, just maybe, we can abandon that fiction and still get all the goodies we think only it can provide.

I've argued that the Cartesian picture contains two components: first, the idea that we each possess a nonphysical mind that interacts with the brain and body, receiving information about the state of the body and the external world, and giving the body orders; second, that the mind when it gives orders does so in a way that circumvents ordinary causal laws. I've also argued that the image projected by contemporary mind science denies both components. The mind *is* the brain—mental life is realized in our brains—and it is subject to natural laws.

Many people will think that the first component, the claim that we possess an incorporeal soul, is less plausible than the idea that we are free agents. They are right, and the reason they are right reveals how it is that, even if incorporeal minds are destined to be dropped altogether from the ontological table of elements, there is something to be preserved in the idea of free agency.

I want to be as clear as possible on what neo-compatibilism commits me to, and how it preserves the ideas that I am morally accountable for what I do and that in some plausible sense I can do other than I do. This can be explained most efficiently by focusing directly on what is involved in being a good person.

There are two main pictures of the good person. One is the picture of the *Good Character,* the second is the picture of the *Principled Actor.*

Advocates of the picture of the Good Character conceive of goodness as emerging from good character, in particular from possessing an interactive set of virtues. A virtue is a disposition to act, probably to perceive and feel, possibly to think, in the right way at the right time in the right situation. Kindness, generosity, fairness, and honesty are examples of virtues. Good people are good insofar as they display the virtues at the right times, in the right situations, and in the right way. Aristotle thought that virtues, once in place, are "permanent and hard to change." (He thought the same of vices.)

The picture of the Principled Actor is somewhat different. Kantians famously espouse the categorical imperative as the right principle to guide action, while utilitarians or consequentialists advocate deploying the principle of utility. The first says to act always and only on rules or maxims that you judge should be consistently and universally applied by any and every moral agent. The second says to act so that you do the greatest good for the greatest number of people in the long run.

Simplifying some, we might say that the first picture is nonintellectualist, whereas the second picture is intellectualist. The person of good character, once she has acquired the relevant set of virtues, doesn't normally need to think deeply about the virtues when she acts. She sees a person in trouble and offers help. The Principled Actor, on the other hand, does consciously consult her principles before she acts. Suppose a Principled Actor sees a way to save many lives by sacrificing one life. Sacrificing one life troubles her, but if she is a consequentialist then she sees, by consulting the principle she is committed to abiding, that it is permissible, perhaps obligatory, to do so.

One reason this is an oversimplification is that a person of good character will sometimes get a mental cramp when faced with certain situations. Suppose she is faced with a situation that seems to call for ruthless honesty. Honesty is a virtue, but ruthlessness, at least normally, is a vice. So she will need to think about things, to reflect before she acts.

Likewise, the Principled Actor will sometimes not have enough time to think about what to do. Forced to choose between skidding a car into a playground filled with children or hitting an elderly pedestrian, she will choose the less bad outcome. Did she consult the principle of utility and consciously compute which course of action would produce the greatest amount of happiness, or in this case the least amount of suffering, for the greatest number of people in the long run? I doubt she had the time.

The advocate of the Principled Actor will hope that the relevant principle makes its way into her blood and bones so that she doesn't need to consciously consult it every time she is faced with a moral choice. Likewise, the advocate of the Good Character will acknowledge that there will be circumstances that produce indecision, what I have called "mental cramps." It will be good, in such situations, for the virtuous person to think about what she ought to do.

Figure 4.4 shows the sort of mind the neo-compatibilist thinks we possess, what I call the picture of a "complex natural mind." I drew a dashed line in the "rationality module" between virtues and habits and conscious deliberation to make room for the idea that principles can go on-line as dispositions, and dispositions can activate thinking about what course of action ought to be undertaken, and what virtues deployed, in a particular case. Furthermore, virtuous people are rarely unaware of the virtues they possess, and if they are this is not a good thing. A modest person is not disqualified from being modest if she knows she is modest. (She is disqualified, of course, if she goes around bragging how modest she is.) Accepting this much we can draw arrows through the open spaces in the line between conscious deliberation and subconscious habits, indicating that thought and habits can indeed interact in the space formerly known as "free will."

DEWEY THOUGHT OF THE MORAL VIRTUES as a subset of habits in general. Once again, he was onto something. An excellent athlete doesn't need to think about every move he or she makes. An athlete who does think about every move is called a beginner. Still, if I am an outstanding tennis player in the U.S. Open Finals, playing, as we say, "on automatic pilot," and I see that my return of serve is not as

FIGURE 4.4 A Complex Natural Mind

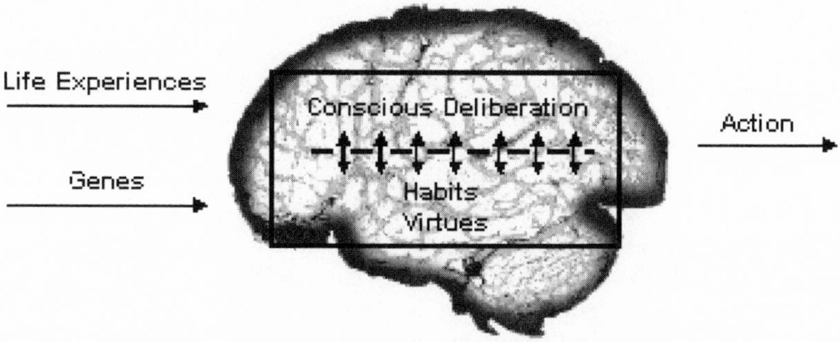

good as usual, I may well think that I should reposition myself in accord with advice my coach has given me for dealing with situations of this sort.

Where Do Virtues and Principles Come From?

Think of the person of good character, who ordinarily acts well without conscious deliberation, as one ideal type, and the person who always consciously routes a moral decision through a principle or set of principles as an ideal type at the other end of the spectrum of moral agency. Here I am using the term *ideal* to mean a useful fiction, not perfection. For what it is worth, I would prefer a world with neither pure type.

In any case, the question we need to ask is where might the relevant moral equipment, the virtues or principles that guide the action of these ideal types, come from?

The Cartesian can—indeed, should—say they are fully self-initiated, completely self-chosen. This is certainly false. By most everyone's lights, an individual needs to learn the right habits of perception, feeling, and action, or to learn the right principles.[8] Evolution may have endowed us with certain good dispositions, basic fellow feeling for example, which can be modified, moderated, and expanded to produce a virtue. Similarly, it may have endowed us with capacities to learn and abide principles. Indeed, Mother Nature undoubtedly did both of these things. But to become a Good Character or a Principled Actor one needs moral education. Where did our

[8]The fact that defenders of the libertarian account of free agency, of "agent causation," just like everyone else, engage in moral educational practices causes a certain instability or tension between what they practice and what they preach. Minimally, they will have to say they are trying to influence their charges. We can ask how much they are trying to influence them and how influencing works. Is influencing just intended to provide information that free will can then use or disregard as it sees fit? Or, if it is intended to do more than that, to have more, or more deeply influential, effects than simply making information available, then how is it to have such effects? Once this question is answered, the opponent of the libertarian will press on, trying to corner the libertarian in a place from which he has no room to maneuver. At which point, if he is honest, he will abandon his ill-fated belief in free will.

ideas of what makes for a good person come from? Thin air is one possibility. Another is that God or nature endowed each of us, as our birthright, with knowledge of what is right and good. A better idea is that we discovered, and are continuing to discover, what makes for a good person and a decent and healthy life.

If this is right, then the causal picture starts to fill out. We detect over historical time that certain ways of being are better than others, and we try various strategies to inculcate the equipment to be good in individual agents.

Do individual agents so trained detect when moral situations arise? *Yes.* Do they engage in self-control? *Yes.* Do they possess powers of self-authorship, powers that once they are trained and in possession of a certain distinctive personality, they utilize in making adjustments to who they are and how they behave? *Yes.* Do they deliberate and reason? *Yes.* Is reasoning and deliberation causally efficacious? *Yes.* Is there any meaning to the idea that a person could have done other than they did and that individuals are responsible for what they do? *Yes.*

The causal network here is immensely complex. But it includes a person with all sorts of agentic powers. I am still here, and how I feel, think, deliberate, and choose makes a difference. Minimally, persons co-author their lives. Indeed, there comes a point in development when we rightly say we are the primary authors of our identity.[9]

No doubt the following objection lurks: "Look, Flanagan, you are trying to trick us. For any agentic capacity $a_1 \ldots a_n$ that you claim exists, you think that it itself is caused. How and on what a person deliberates is to be explicated in terms of natural causes. Furthermore,

[9]One thing I haven't emphasized so far, but will say more about in the next three chapters, is that free action, especially the components that have to do with self-expression, crucially involves what Harry Frankfurt calls "identification." I want my voluntary actions to be more than expressions of what I want now. I want my choices, my work choices, my aesthetic judgments and choices, and perhaps most importantly my moral choices to reflect what sort of person I see myself as, and to a certain extent to reflect what sort of person I want to be—assuming I want to keep getting better and better. Often I have desires I don't consider good or worthy, and sometimes, especially when such desires are persistent, I work not to have these desires. In such cases I work from some image of the sort of self I think it best to be, from a self-image that I have come to view as worthy of identification.

you don't really think that any act actually done could have been different. And if you don't think that, an agent is never responsible in our sense of responsible. And if an agent is never responsible, then the entire rationale for our moral and legal practices of reward and punishment is dissolved."

The critic is at least partly right. What the neo-compatibilist means when she says that an individual could have done other than she in fact did is that *if* that person had seen the situation more clearly, had been sensitive to reasons she was not in fact sensitive to, she could have done otherwise. She does not mean that the person could have acted other than she did. If she acted from deterministic rational causes, whatever they were, then these necessitated her act. If some indeterministic neural firings caused her to think (mistakenly or not) that "This is a really good idea, I'll do it," her act was also necessary.

The fact remains that according to the account on offer, agents can in fact normally do any number of things. When I consider a number of options—going to the movies, having a friend over for a visit, staying home and reading a book—I normally do so only when all the options are open to me, when all are possible and, to some extent, attractive. When I deliberate and choose what to do all three options are open to me. If I were to choose any one of them, nothing would prevent me from carrying through on that choice. Insofar as the worry is about what I can do, the neo-compatibilist can make clear sense of the concept of live options. Furthermore, she can make sense of "could have done otherwise" in the following sense. Even after I choose, say, to go to the movies, it is still true that I could have stayed home and read had I chosen to do so.

Furthermore, what the neo-compatibilist means when she says that a person is responsible for some act is, first, that the act was routed through the conscious deliberation/habit module; second, that that module is adjustable from the inside, by the agent, and from the outside, by way of feedback from the moral community; and third, that by virtue of being routed through a modifiable cognitive module, the person can learn to respond differently the next time. She is responsible in the sense that she has the capacities to respond differently in the future. We might say that a neo-compatibilist agent is responsable (with an "a", not responsible, with an "i") in that she is *able* to modify her future actions in light of her own and the community's responses to her past actions.

If the conceptual model, the rationale, required by our moral and legal practices of reward and punishment is the model of responsibility with an "i", where the agent could have done other than she in fact did given her exact state of mind at the time, then it is true that the neo-compatibilist picture dissolves that rationale. To what degree the dissolution of the rationale requires changes in our actual practices I leave as an open question. Perhaps the situation is akin to Christmas without Santa Claus: Presents are nonetheless exchanged. The practices stay in place because the fiction that seemed to motivate them, or to be required to make sense of them, turns out to have been inessential.

Where Are We?

I've been suggesting a way of preserving, within the scientific image, the idea that we are agents who perform voluntary actions. The picture I am recommending preserves some, possibly a lot, of what ordinary talk of free will entails, and it provides a rationale for moral inquiry and moral education. I like to think that this allows a defender of the manifest image some space to move, some space to adjust her picture so that the manifest and scientific images are not so starkly inconsistent.

To a resolute defender of the manifest image I put this challenge: Explain to me what it is you are doing when you engage in moral education. Why do you attempt to influence the young to learn what is right and good? Where do your theories of what is right and good come from? Why is it that some people deliberate well and others badly or not at all? Why is it that indigency, poverty, and discrimination are so powerfully connected with lives that go awry, with people who do bad things? If free will is a prime mover unmoved, why are you attempting to move it?

The answers, I know from experience, will focus on influences and resist talk of determinants. This is fair enough. Our moral educational practices, and poverty and discrimination as well, are only influences. They are never the complete causal determinants of the state of a person's mind or of how she acts. But the defender of the scientific image will claim that there is always some set of causes (including perhaps some indeterministic ones), some set of influences, that taken together is the total cause. And once again the defender of the manifest image is set the task of defending a conception of agent causation that has never been ably defended.

It deserves mention that the defender of an image that includes agent causation is quite a bit worse off than the defender of an image that includes God as creator. The cosmological story that begins with the Big Bang does not, as I have said, explain how the singularity that banged got there in the first place. This leaves room to claim that once, at least, an incorporeal agent caused the physical universe to come into being. Personally, I think this is a bad idea, because it is completely without supporting evidence. But the theist at least has some free space to play with when it comes to the origin of the universe. No such space exists once we are *inside* the universe. There is no workable conception of agent causation that has ever been articulated and defended. Furthermore, the fact that we are not in possession of strict causal laws for all aspects of the physical world, and thus can never specify the total and complete cause of any event, is best explained as sometimes an epistemic problem and sometimes as an ontological one. We are finite creatures with finite cognitive capacities. Given that we are very intelligent but not omniscient, and given that the natural world may well include both deterministic and indeterministic causes, the goal of making the world completely intelligible in natural terms may well eternally elude us.

The regulative idea that this world, however it got here, is fully natural, obedient at every juncture to whatever laws nature abides, has proved again and again to be progressive, to yield knowledge. The regulative ideal that holds out for the sort of causation required of free will has led nowhere.

The view that assumes nonnatural causation of the sort a Cartesian free will requires not only assumes something we have good reason to believe is false and is lacking in credible resources to explain the advances of the human sciences, but is actually a morally harmful picture. It engenders a certain passivity in the face of social problems that lead certain individuals to be malformed. There are bad people in this world. But if we think that bad people are bad simply, or even mainly, because they choose to use their free will badly, we are making a big and costly mistake.

One Last Disquieting Suggestion: Moral Epiphenomenalism

So far in this chapter, I have been trying to find grounds for conciliation between the manifest and scientific images. I've claimed that when it comes to the nature of mind (as a nonphysical substance) and free will that the two images are inconsistent. But I've been searching for some common ground in an effort, as we say, "to make nice." At least I was trying to make nice until a moment ago, when I implied that staunch defenders of the Cartesian conception, in addition to holding an incoherent view, hold a position that can be morally harmful. I think this, so I'll let it be.[10]

In any case, I want to close by pointing to a set of results from social and personality psychology that undermines, to some extent, the pretty neo-compatibilist picture of agency I have been advocating. (This picture, recall, is diagrammed in Fig. 4.4.)

The basic idea of the pretty picture is to highlight how in normal cases of moral agency, principles and virtues are the proximate causes of action. Moral action is routed through well-entrenched, identity-constitutive virtues and principles that the good person abides.

I'll suppose that everyone reading this book is nice enough. I bet you see yourself as a Good Character or a Principled Actor, likely as both. So imagine yourself in this situation: You are shopping at the mall and need to make a phone call. After you complete your call and

[10]There is an interesting reply that has been tried by many very smart people ranging from Blaise Pascal to Fyodor Dostoyevsky to William James, which involves the idea that when all the pros and cons of believing in things for which there is no evidence are added up it *is better,* in the sense that a life will go better, if you believe in certain things that are false, or, what is different, in things for which there is *no* evidence. Believing in God and free will tops the lists here. Members of the Kansas Board of Education belong to this group. They are on record asserting that if the Big Bang theory displaces the belief in divine creation and Darwinian theory displaces the belief that humans have souls and free will, life will lose all meaning and purpose and that mankind, by having lost its foundation, will go to hell.

leave the phone booth, a young woman drops a folder full of papers in your path.

What should you do? What will you do? Well, we all know you should stop and help. I also know that almost everyone will predict that they will stop and help—we are not talking about a stranded motorist who needs you to rebuild her engine, or even to change her tire.

But in fact the probability that you will help is low. In one study, only one-third of the people exiting the phone booth helped pick up even one sheet of paper. Two-thirds went on their merry way.

Now the sly foxes who performed this experiment noticed a funny fact about telephone callers in public booths—they always check to see if any change is in the change return before exiting. So the researchers applied for a federal grant that required, in equipment, 16 dimes. That is one dollar and sixty cents in 1973 currency. Aided by the young woman with the folder filled with papers, the experimenters placed these dimes randomly in 16 phone booths throughout the mall. So 16 people found a dime, and 25 came up empty-handed. Guess what? Fourteen out of 16 who found a dime helped, and 1 out of the 25 who didn't find a dime helped. That is, 87.5 percent of the lucky ones helped, and 4 percent of the others.

Yes, I know a dime could hardly affect *your* behavior, as you are a person of good character and principle. I also know that you will predict that neither you nor any of your friends would abide Stanley Milgram's infamous obedience experiment in which he paid subjects to help a person learn by shocking them for wrong answers. Milgram studied over one hundred subjects over a three-year period, between 1960 and 1963, and found that a full two-thirds gave shocks at the highest level, 450 volts, a level that was clearly marked as VERY DANGEROUS, to a fake learner who had started screaming and pounding the walls at 300 volts.

Individuals insist that they would never go this far. Indeed, almost everyone thinks they would stop shocking the learner at the first sign of discomfort. When told of the Milgram results and shown pictures of various individuals, good looks and being a female are judged to be good predictors of who will not comply. But this is wrong. There is no discernible difference in looks, gender, race, socioeconomic status, educational level, country of origin, or character traits between maximally compliant persons and maximally rebellious ones. In the best social psychology text written, by the late Roger Brown of Harvard, Milgram's compliance rate is described as "one great unchanging result." Phil Zimbardo of Stanford thinks that "the reason we can be so readily manipulated [in Milgram-like experiments] is precisely because we maintain an illusion of personal invulnerability and personal control." The manifest image of free will is part of the problem, not the solution.

Now neither the phone booth experiments nor Milgram's experiments can be done nowadays. The latter because they have been judged unethical; the former because phone calls from malls are now made on cell phones, which lack change return slots and permit their user to talk and leave footprints on the fallen papers at the same time.

In any case, hundreds of experiments by social psychologists show that innocuous causes can have powerful effects on behavior. Tell seminarians who have just prepared a sermon on the parable of "The Good Samaritan" that they need to hurry to get to another building, and they will not help a person in terrible distress along the way. Watch a person wait for the liquor store attendant to go to the storeroom and then, with bravado and clear intent, steal a case of beer, and a full 80 percent of bystanders will not say a word when the attendant returns. If the attendant asks suspiciously where the other guy went, a full 40 percent still do nothing to help him.

When I first suggested that these results had important implications for our conception of moral agency, in my 1991 book, *Varieties of Moral Personality*, several fellow philosophers ran with the ball. Gilbert Harman, a distinguished philosopher at Princeton, has even claimed that these results show that there are no such things as character traits. Indeed, he titled a recent paper "The Nonexistence of Character Traits." Remember the Ptolemaic who said, after losing to the Copernican, "There is no earth."

Most of the response to my suggestion has focused on its implications for virtue or character ethics. I remain convinced that there are character traits, although they are hardly as constant or resilient as some think. Aristotle is often treated as the chief culprit here. But Aristotle understood that the virtues may not steel a person against all countervailing forces. In speaking of the courageous person, Aristotle acknowledges that "some dangers are terrible beyond human strength." In Euripides' *Hecuba,* Hecuba is introduced as a person of great and consistent virtue. She has maintained her courage, dignity, and nobility despite losing her city, her husband, and most of her children. She even withstands the sacrificial offering by Odysseus of one of her two remaining children, Polyxena, to appease Achilles. But when her trusted friend Polymestor kills her last child, Polydorous, Hecuba comes undone. She becomes a nihilistic, vengeful person—a murderer of innocent children herself.

If social psychology creates problems for virtue ethics it is not because it shows that there are no character traits. It is because it shows that these traits can be overridden, circumvented, or made to go off-line even by innocuous, not remotely "terrible" causes.

Insofar as work in social psychology and personality psychology has implications for our picture of moral agency, it affects the pictures of both the Good Character and the Principled Actor. People whom we judge as kind to the core *and* people who abide the categorical imperative and would never treat another as a mere means to an end behave no differently in Milgram-like situations.

Furthermore, you may be as committed as you please to some universal ethical principle. The best predictor of whether you will resist Nazi atrocities, to pick an example that has been well studied, is not the degree or depth of your commitment to that principle. It is whether you score high on an adventurousness, risk-taking scale, and whether you have a sense of being socially marginal. A morally good parent also figures favorably in the mix, but not as the most important factor.

You might think that if you yourself were the object of horrible social discrimination, you should look for help among people who were picked on in middle school, and who favor riding motorcycles without helmets, over those who have read the Bible, Aristotle, Kant, or Mill. You would not be wrong in thinking this.

Here's the point. Recall the picture of epiphenomenal agency in Fig. 4.3. The basic idea behind epiphenomenalism, I said, was that the factors we think are playing a dominant role in the causation of action are not, in fact, playing that role. I worked hard to convince you that there was a picture of moral agency—the neo-compatibilist picture—that could fully accommodate the scientific image and much of the manifest image. And I suggested that we should learn to live with that picture of agency since it eliminates outright inconsistency between the two images and is the best we can do given the current state of knowledge.

I still think this. But the work in social and personality psychology I have been discussing suggests that even if we can make room for the picture of agency I recommend, according to which we act and/or deliberate in accordance with the virtues we possess and principles we avow, we do not always act this way.

To be sure, we will believe that our character and principles consistently guide us to do what is right and good. But the fact that we believe this hardly makes it so. Small, difficult-to-detect factors can have profound effects on how we think, indeed on whether we think at all, and how we act. So there will be glitches and misfires in being good, and causal forces that we do not see may carry the day, even as we sincerely believe that our character and principles are having their way. In its industrial-strength version, epiphenomenalism says that our conscious will module is always out of the loop. What I am suggesting is milder: Certain results in social and personality psychology should make us worry that the epiphenomenalist suspicion is at least sometimes legitimate.

I don't see that this is completely bad news. As they say on the street, "knowing is half the battle." If I know that subtle situational variables can have a profound effect on how I behave, I can learn to keep my eye out for them. Epiphenomenalism as a general thesis is almost certainly false. But knowing that our conscious mind exerts less power than we think it does can empower us, and at the same time can help to undermine two related myths that we are sometimes gripped by.

The first myth is that we are in complete control of what we do thanks to our possession of a completely unconstrained will. I have devoted most of this chapter to exposing that myth for what it is, while maintaining that there is nothing in the scientific image that eliminates any of the features that a credible conception of free agency must allow. If the argument of this chapter has succeeded, then the entire list of requirements with which we began has been shown to be not just possible, but actual for creatures with our sort of complex natural minds. Every requirement is met. Here is how the scorecard looks:

- *Self-control*
- *Self-expression*
- *Individuality*
- *Reasons-sensitivity*
- *Rational deliberation*
- *Rational accountability*
- *Moral accountability*
- *The capacity to do otherwise*
- *Unpredictability*
- *Political freedom*

That accomplished, there is a second myth that needs to be dispelled, and raising the epiphenomenalist suspicion helps to expose and undermine it. The myth I have in mind is that we are capable of complete rationality. Like its radical partner, which says we have completely unconstrained wills, this is a silly, misguided idea. Recall that we are animals. Our minds are hardly perfectly designed, and all manner of external and internal quirky contingencies are capable of influencing the ways we feel, think, choose, and deliberate. We are creatures with capacities to think and act voluntarily and we are creatures who reason. But we can only exercise these capacities in ways suited to the sort of incredibly smart, but not perfectly self-controlling or perfectly rational animals that we are. Our practices of holding people morally and rationally accountable will need to pay close attention to the many forces that constrain our choice and our reason. By so doing, we will show due respect for our increasing knowledge of human nature and perhaps discover more humane ways of responding to and treating our fellows.

The scientific image itself is continuously subject to refinement and change. Each time science speaks about the nature of persons, it adjusts its view of what a person is and what makes people tick. When good science speaks, we had better listen, even if listening requires changes in the manifest image of persons. The dialectic of settling into an image, having it challenged, and making adjustments is our eternal lot. Who and what we are, what our nature really is, is something we are still a long way from knowing in detail.

FIVE

Permanent Persons

> I don't want to achieve immortality through my work. I want to achieve immortality through not dying.
>
> —WOODY ALLEN

> Where death is I am not; where I am death is not, so we never meet.
>
> —EPICURUS

NOTHING IS MORE IMPORTANT, at least to me and to my friends and loved ones, than who I am. But what is this me, this self that is so important? Suppose there is a self, suppose the world is filled with selves, where and what are they?

It is part of the manifest image in the West and in most parts of the East that whatever else the self is, it is that part of me that has permanency and constancy, and that explains what really makes me who I am, what really makes me, me. Buddhism is pretty much unique in questioning this picture. According to Buddhism, the idea of a permanent, constant self is an illusion, and a morally dangerous one. The last words of Siddhartha Gautama, the first enlightened one, the first Buddha, were: "Impermanent are all created things: Strive on with awareness." This is, I think, the truth and good advice as well.

My son, Ben, and I visited Dharamsala, India, for a week in March 2000 to meet with His Holiness, the 14th Dalai Lama, who lives there in exile from his home in Tibet, having been driven out by the genocidal conquest of Tibet by China. Eventually, I will have some things to say about why Buddhists think that there is no such thing as *the self,* as well as why they recommend quietism about theological issues. For now I will simply repeat that this is a rare view. But the Buddhist view fits better with the scientific image than its manifest relations.

My own view is that there is no self, if by *self* we mean some permanent, immutable, abiding essence that determines who we really are and that accompanies us on our ride through life (and possibly after we die). What we call "the self" is an abstract theoretical entity. The self is, as Daniel Dennett puts it, the "center of narrative gravity." The projects of self-knowledge and self-location do result in the discovery of certain patterns of thought, memory, and behavior—patterns that are, in some genuine sense, abiding. What we think of as our unique individual selves consist of the integrated set of traits we reliably express and embody, the dispositions of feeling, thought, and behavior we reliably display, as well as a certain kind of psychological continuity and connectedness that accrues to embodied beings by virtue of being-in-the world over time. When we think or talk about ourselves, our aim is to describe these patterns.

Furthermore, because each of us is a continuous individual organism attached uniquely to our own nervous system, we have a unique first-person sense of ourselves. But if we look closely, it is a sense of our self as a complex and relatively continuous entity, not as something permanent and simple, and certainly not guaranteed never to change.

My Self, My Soul

One familiar view is that the self is the soul, what the Hebrews called *nepesh*. What we are trying to get at and locate when we talk of "the self" is our soul. In the standard view, a soul comprises a person's essence and is not itself part of the natural fabric of things. Philosophers usually refer to souls as "Cartesian egos" since Descartes claimed to *prove* that humans are possessed of souls. Thus I'll use the terms *soul* and *Cartesian ego* interchangeably. Notice that I am not identifying Cartesian *res cogitans*, my immaterial thinking stuff, with my soul, with *nepesh*. Descartes, in fact, often speaks of the "mind" and "soul" as equivalent, but there is also available in his writings the idea that one's mind is backed up or made possible by the possession of something additional to it. It is this extra ingredient that I am calling "soul" or the "Cartesian ego."

The reason for imputing that each individual, in addition to being a complex of an immaterial mind and a physical body, possesses a soul emerges from asking this question: Is Cartesian dualism sufficient by itself to account for personal identity through time, and possibly for personal immortality? Descartes seemed to think the answer on both counts was "yes." But several important post-Cartesian thinkers have disagreed. Although Descartes did say that my mind (soul) could exist without my body, these thinkers felt he didn't say enough to explain how in this life my mind makes me the person I am, and how his idea of mind explains how my essence is preserved in an afterlife.

What Descartes's philosophy was missing, to put the matter in stark terms, was a way of seeing how the mind, despite being my essence, had the sort of nature that could really carry and preserve a permanent and abiding essence. The life of a Cartesian person is the life of a mind and body travelling together down life's road at a particular place and time. In spite of its magical powers of free will and its essential separateness from body, the life of such a mind could be conceived in terms of its unique career, as constituted by its experiences, actions, and particular personality. All this seems largely contingent. Plunk my incorporeal mind in a different body at a different time and its history would have been different. If there is something that is really me, really my essence, it should not be simply reducible to its career while embodied. My essence might make my contingent experiences possible and it might, as it were, hold my character or personality as expressed or revealed in my lifetime, but still it seems as if my essence should be something more than just my embodied mind's life history.

What might that be? Two possibilities stand out. Either the mind has a pure immutable extra ingredient, its essence, that was insufficiently articulated by Descartes and that constitutes something like its pure, permanent, structural form—something it possesses, indeed that it is, prior to getting entangled with the world and that can survive that entanglement after death. Or, a less parsimonious idea is that there is, in addition to mind, a soul—a third, extra ingredient that contains the incorporeal mind and that possesses the properties of indivisibility and immutability. The first view tells us we need to elaborate the Cartesian conception of a nonphysical mind to make it reveal the mind's essence, the part of *res cogitans* that is the Cartesian ego, or *nepesh*. The second view accomplishes pretty much the same thing but in a less elegant way. It requires that we think of persons as having a tripartite structure: We have a soul and a mind and a body. Its inelegance notwithstanding, this second view arguably makes it easier to say what makes a person identical over time. What makes her the same person is her possession of the same soul. This view also helps with certain puzzles that beguile philosophical theology as it tries to explain in what sense "I" might continue to exist after I die, and, in particular, exactly what part of me is preserved in an afterlife. Possibly it is my pure essence, my soul that is preserved, but not necessarily my mind insofar as it carries the vagaries of my distinctive personality, memories, and experiences.

Why Discuss the Soul?

Given the naturalistic accounts I have offered of the Cartesian ideas of an immaterial mind and free will, why should I complicate matters by discussing an even more dubious idea, the idea of the soul? But in fact I must discuss the soul, and its close mate "the self." These ideas are part of the manifest image. They figure centrally in discussions within perennial philosophy of the topic of personal identity and in discussions of personal immortality.

The belief in souls—or at least in some permanent and abiding part of myself that makes me me, and which in addition may survive my death—is widespread. According to a host of polls in the last decade, somewhere between 70 and 96 percent of Americans believe humans possess nonphysical souls, and they believe that this soul continues to exist forever after the body dies. These numbers are roughly the same as they were a century ago. No doubt many people think that having a Cartesian mind is enough to make a soul. But as we just saw, the perennial philosophy whose job it is to refine and endorse the manifest image is not in unanimous agreement that Cartesian dualism is sufficient to answer questions about either personal identity or personal immortality. So we need to examine what work the concepts of the self and the soul might do that cannot be done by Descartes's dualistic picture or the naturalistic view of persons I have promoted thus far.

There are three distinct but related issues that lead to the positing of a soul, a self, an abiding ego.

1: *The Possibility Conditions for the Unity of Experience*: How is experience possible? One answer we have already examined involved positing a mental faculty that makes experience possible, either Descartes's *res cogitans* if one is a dualist, or the brain if one is a naturalist. Either answer might be viewed as sufficient for explaining how individual experiences are possible, but they fail miserably when it comes to explaining how it is possible for experiences to be unified and experienced as belonging to a single subject. We appear to need to posit an ego, an "I" that has or contains experience, to be the site where experiences come together. This is the self or soul.

2: *Personal Identity*. I seem to be the same person over the course of my life. My experiences come and go, but "I" stand in the same relation to my experiences over time, this making them my experiences. Philosophers ask what are the logically necessary and sufficient conditions for a person P_2 at time t_2 being the same person as a person P_1 at some earlier time t_1? What makes me, Owen Flanagan at 52, the same person I was at 42, or 32, or 22, or 12, or 2? My body doesn't seem like the right answer, for it has changed a lot. My mind doesn't seem like the right answer either, for if my mind is characterized as that which has and houses my memories, experiences, and personality, it too has changed a lot. Perhaps there is a pure ego that is part of my mind, or even a third ingredient beyond my body and mind, my transcendental ego, which stays the same amid the flux and makes me the same person over time.

3: *Personal Immortality*. Most people think that I can survive my death, or at least that this is possible. My body doesn't seem like the right sort of thing to survive my death—indeed it is the very thing that dies. Perhaps my mind, considered as a nonphysical substance and as the conscious stream of experience and memories that also houses my personality, survives. But some sacred scriptures describe the afterlife in a way that involves purified survival. I don't go to heaven still remembering all the experiences I had. Nor do I need the equipment that allowed me to have such experiences. It is not clear that I even pass through the heavenly gates with all my memories, or with my personality intact. There is an ancient myth according to which the soul after death crosses a river, which cleanses it of earthly memories. One idea of survival is that my soul is purified in a way that unifies my rational and moral soul with God, who embodies Reason and Goodness as such.[1] So perhaps what survives death is the part of me that is most pure, permanent, and, somewhat ironically, the least personal, but which nonetheless is what contained and made possible the personal life I had on earth. My true self, my abiding ego, my soul, my *nous*, my *nepesh*.[2]

The Philosopher as Gadfly

The belief in the soul is, like many things we believe in, normally accepted without argument, let alone proof. Philosophy, however, is in the business of examining unexamined beliefs. Some, possibly

---

[1]Our ideas about immortality are inchoate. Plato thought that *nous*, our impersonal rational and moral soul, preexisted the body and survived after it. Aquinas, following Aristotle, seems sometimes, at least, to think of immortality as involving something like the survival of my *nous*, but not my actual person or personality. Most people I ask hope that they survive in some beatified way that still preserves them as the person they are. Perhaps this desire is just a symptom of modern individualism run amok. But pretty much everyone sees, upon a moment's reflection, that preserving all of their identity will not be the best solution. They might if they were a gourmand still want fine wine and *pâté de foie gras*, and they don't serve those in heaven. Plus it would be inconvenient and messy if in heaven one faced, say, all one's past lovers, not to mention enemies, and came armed with all the old feelings for these people.

[2]The idea of impersonal salvation is available within Hinduism and Buddhism. *Nirvana* is a state akin to what we in the West might think of as non-Being, where once one has achieved enlightenment one is free to no longer exist. The doctrine of a cycle of rebirths and reincarnations that are normally required before one achieves *nirvana* was only proposed in the eighth-century C.E. and then spread like wildfire among Hindus and, to a significant but lesser degree, among Buddhists. Furthermore, it is worth noting that each reincarnation involves forgetfulness. And even if my (bad) *karma* causes me in my next life to live as a rat, the rat presumably not only lacks my memories, it also lacks my personality.

many, of these beliefs will pass muster. The vast number of things we believe, we believe on the basis of authority. This is a good strategy because it gives us each a jump start in life. We are born into a position to be taught, to trust, and thus to accept the wisdom of the ages. For the most part, we check almost none of the answers provided as our birthright. As Baruch Spinoza pointed out, even knowledge of our birth date and of who our parents are is based on hearsay. Things are as our ancestors, themselves recipients of the wisdom of the ages, say they are. Not to trust what they say and believe would involve having to reinvent the world.

In general, we are epistemically conservative, checking past wisdom only when it causes trouble. Scientists, according to Thomas Kuhn, are like everyone else in that they assume, but don't recheck, the wisdom of the ages until accepting some set of "received truths" starts to cause trouble. Normally it takes a lot of trouble, numerous anomalies, to get a scientific community off its duff and on the way to entertaining a new paradigm.

Socrates believed his society was hypocritical and complacent. Athens claimed to be democratic, free, and open to moral and political challenges to its traditions. But in fact, Athens was closed to scrutiny of its received ways of thinking and being. Socrates compared his mission to that of a gadfly that spends its life bothering a lazy horse. His job, as he saw it, was to question unquestioned assumptions in order to see what, if any, basis they had. He was not interested in questioning every unexamined belief. His targets were invariably beliefs that were important and that appeared to lack rational support—ones that he saw as widely held but unable to withstand scrutiny. The belief that those in power were wise simply by virtue of being in power was one of his targets, as was the alleged concern of the elders for the moral improvement of youth, which he saw as a self-serving effort to create obedient moral and political allies.

My primary target as an heir to Socrates's project is the widespread belief in our permanency as persons, the belief that there is an abiding "I" that accompanies experience but is irreducible to the continuity of our natural lives as embodied beings.

Still, if I meant what I said about epistemic conservatism, why am I questioning these beliefs? Perhaps they are false, but they aren't causing trouble. The answer is that they *are* causing trouble. Most philosophers and scientists in the twenty-first century see their job as making the world safe for a fully naturalistic view of things. The beliefs in nonnatural properties of persons, indeed of any nonnatural thing, including—yes—God, stand in the way of understanding our natures truthfully and locating what makes life meaningful in a nonillusory way.

The scientific image is incompatible with viewing ourselves in the way many, possibly most, do. Furthermore, historical evidence abounds that sectarian religious beliefs not only lack rational evidence or support, but that they are at least partly at the root of terrible human practices—religious wars, terrorism, and torture. Yes, I know the answer; such calamities come at the hands of fanatics. Even if this is true, the fact is that fanatics are fanatics because they believe that what they believe is indubitably true.[3] So the belief in the truth of certain beliefs is part of the problem. And again it is the philosopher's job to examine the grounds of beliefs, and thus to assess their justification and ultimately their truth.

Some acquaintances—I run with an odd crowd—think that the belief in souls is extremely silly, but fortunately also extremely rare. They are wrong about this. Belief in a soul is not even an idiosyncrasy of Westerners. The belief in a nonphysical soul that comprises an individual's essence, that is not physical, and that survives bodily death is a feature of virtually all of the world's great religions.[4]

That said, I want to post some cautions. Polls that ask people what they believe, that ask them to agree or disagree with certain sentences

[3]In January of 2001, I read a long article about the Al Qaeda terrorist group led by Osama bin-Laden. Several former soldiers in his army expressed deep disappointment that they had not died, as they had hoped, in the holy war that bin-Laden has declared. Their reason: Martyrdom is the most noble way to die and if one dies as a martyr in a jihad one gets a straight ticket to heaven and spends all eternity in the company of virgins.

[4]There is some discrepancy between the views of ordinary intelligent people and some theologians. A few contemporary liberal theologians recommend thinking of persons along lines suggested by Aristotle and Aquinas according to which we do not possess any nonphysical components. And some very liberal theologians even think of the quest to be in relation with God for all eternity as a simple, benign expression of the human thirst for goodness and enlightenment.

affirming or denying some metaphysical or theological belief, are insensitive to subtleties. We get little information about the nature or object of the belief or about its degree or depth. In the spring of 2000, I attended my 18-year-old son Ben's graduation from Carolina Friends School. The graduation took the form of a traditional Quaker meeting. We were asked to sit silently, to let ourselves be with our thoughts and feelings, disappointments, pride, and hopes, and to speak "only if God moved us to speak." Quakers speak of "God," but theirs is an unusual God. In my experience, God-talk among Quakers is a way of speaking that allows each individual's interpretation of the meaning of the term "God." Insofar as Quakers have anything remotely like a theology, "God" means something like "that which represents good, and love, is larger than us, dwells among us, and can be amplified through us and by us." But you will not get into a disagreement with a Quaker if you say that what you mean by "God" is all of creation, this being larger than each of us individually, and that which represents the forces of good and love. This way of speaking is pagan in certain respects. But the reason the Quaker will not disagree with you is not just because she is a "Friend" and thus committed to pacifism in thought, speech, and action. Her religion is in fact pretty noncommittal, intentionally vague when it comes to "God's nature." Talking and thinking about "God" is for Quakers a way of focusing the mind on the nature of goodness in the world.

My former wife, Joyce, and I were seated together at the graduation. Joyce is an active member of a Unitarian Church. She would probably answer a poll that asked if she believed in God by saying "yes." But again I know that what she means by "God" is not what a Roman Catholic or Jehovah's Witness means.

The polls that reveal that astounding percentages of Americans believe in God, angels, souls, and an afterlife are blind to this nuance. There is a fair amount of unclarity about the precise nature or object of belief, and about the degree of commitment to it (more information would be gathered if people were asked to rate their degree of commitment on a 1 to 10 scale rather than simply say "yes" or "no"). Furthermore, people are not asked whether they think their beliefs are true. One might think this is redundant, since when we say we believe some proposition, what we mean is that we believe it to be true. In my experience, metaphysical and theological beliefs are atypical. People I speak with sometimes acknowledge that they believe in the soul or in a certain image of God and then acknowledge that they are not sure if what they believe is true.

Often, they will say that they are in no position to argue the truth of what they believe. My sister Nancy recently reminded me that as kids we were taught that faith was defined as believing what one could not show to be true. If this is the definition of faith, faith is something the philosopher will have trouble with. Knowledge is warranted belief, belief for which at least some reasonable evidence can be produced. As we sometimes say, knowledge is justified true belief. If faith is believing something for which no evidence can be produced, then it is an odd thing for people to admire. It might be better to put the point this way: It is not faith as such that the philosopher will have trouble with. I have faith that I will finish this book, and I have faith that my children have what it takes to live good lives. But in these cases, I have faith in things for which there are good reasons to believe. The worry is that in the theological case, there just may be no good reason to have faith.

In any case, people like my ex-wife, Joyce, and our Unitarian and Quaker friends will sometimes acknowledge a belief in God, but go on to focus not on the truth of the belief but on the way the word "God" functions in their community. It sets people's sights on possibilities for making the world better. Some—my Quaker and Unitarian friends especially, but some Catholic and Jewish friends as well—even acknowledge that it doesn't matter if what they believe is true.

But just as "God" talk, religious ritual, and music serve a purpose for my Quaker and Unitarian friends that is unrelated to the factual truth of various assertions, religious music lets me feel the force of beauty and goodness and love in a way rock and roll never does. The fact that only religious music, where there is explicit talk of eternal souls and of God, takes me to this place where I sometimes want and perhaps need to go does not require me to believe in the literal truth of what is being sung. Normally, I don't even think about truth and falsity.

The project of showing that beliefs in an immortal soul and in God are irrational should not be taken as denying that the expression of religious feelings and thoughts may have many good functions.

Talk of the supernatural involves an attempt to get at something important—that which is larger than us, awesome, and thus awe-inspiring—and talking (or singing) about the supernatural can evoke unusual and worthwhile emotional and ethical attitudes. Nonetheless, with these caveats on the table, it still seems pretty clear that many people believe that it is true that there is a God, a God who is non-physical, typically has the role of creator, and who in addition embodies and represents the forces of good. Many people—most Americans, at any rate—also believe that humans are truly possessed of a nonphysical soul that outlives their body.

Although I am respectful and appreciative of philosophical and religious communities that use certain words not so much to express truths but to move people to be better, my aim is to examine beliefs that are believed true. This is the philosopher's job. It is an interesting and important question whether beliefs that are not known or even thought to be true; or that are expressed with acknowledgment that evidence and argument for them cannot be provided; or that are thought to be true but are in fact false, can be good for individuals or communities.

For myself, it does not seem remotely incredible that thoughts of something that is all-good, all-loving, as well as immeasurably greater than us—despite the distinct possibility that there is nothing person-like that fits the description—might inspire certain people to be better than they would otherwise be. That said, there are many great thinkers who have thought that the metaphysical, theological, and ethical views fostered by religion are not only false but, all things considered, the cause of much that is evil. David Hume, Karl Marx, Friedrich Engels would agree with the Nobel Prize-winning contemporary physicist Steven Weinberg when he writes: "The prestige of religion seems today to derive from what people take to be its moral influence. . . . I think on balance the moral influence of religion has been awful."

Truth, of course, is not the only thing of value. What is good, beautiful, even fun, is also valuable. And these values can compete with truth. It is fun to believe in Santa Claus, even though Santa does not exist. When the positive effects—amusement, goodness, or beauty—are such that they can override the fact that a belief is false, is a tough call.

Even if beliefs in the soul and a divine creator can be defended as serving a positive function, we will still want to know if there is good evidence for them or little evidence, or whether at the limit they are simply illusions. As I have said, my mission is to work that terrain—to examine whether the belief in an immortal soul, a Cartesian ego, that exists in addition to my organic self is justified. Examining this question also requires examining the warrant for believing in God, the paradigm Permanent Person. One common reason people seek to make sense of the self in a way that deems it permanent and immutable is that only these characteristics offer plausible prospects of eternal communion with the divine. The problems of what the self is, of personal identity over time, of the prospects for an eternal nonearthly afterlife, and of God's existence reside in close conceptual proximity within the dominant manifest image. Sorting out the status of one thus requires attending to the others. The last belief, the belief in a personal God, receives the same verdict as the belief in a permanent, immutable soul: There is no reason to believe it true. This conclusion does not answer all questions about whether this belief or the whole set taken together, even if false, might have some positive personal and communal functions. For now, I'll claim agnosticism on that issue.

Disarming the Body

Why would the belief in the soul continue to be so compelling, despite the serious disadvantages it creates for those who desire a conceptually coherent worldview? One answer is that the incorporeal soul provides a basis for our intuitions about the conscious self.

A favorite, but gory, thought experiment deployed by philosophers who think humans have an abiding essence involves imagining amputations. Descartes tried this tactic, as did the Scottish "common sense" philosopher Thomas Reid in the eighteenth century. Descartes insists that one's essence, one's mind or soul is indivisible, it has no parts. In the sixth of his *Meditations,* he writes:

> [W]hen I consider the mind, that is to say myself as I am only a thinking thing, I cannot distinguish any parts, but apprehend myself to be clearly one and entire; and although the whole

> mind seems to be united to the whole body, yet if a foot or an arm, or some other part, is separated from my body, I am aware that nothing has been taken away from my mind.

A few years earlier, in the *Discourse on Method,* Descartes had argued that each of us can know for certain that he is a

> substance, the whole essence or nature of which is to think, and that for its existence there is no need for any place; nor does it depend on any material thing; so that this "me," that is to say, the soul by which I am what I am, is entirely distinct from my body . . . and even if body were not, the soul would not cease to be what it is.

In a similar vein, Reid writes:

> A part of a person is a manifest absurdity. When a man loses his estate, his health, his strength, he is still the same person, and has lost nothing of his personality. If he has a leg or an arm cut off, he is the same person he was before. The amputated member is no part of his person, otherwise it would have a right to his estate, and be liable for part of his engagements. It would be entitled to a share of his merit and demerit, which is manifestly absurd. A person is something indivisible. . . . My thoughts, and actions, and feelings change every moment; they have no continued, but a successive existence; but that self or I, to which they belong is permanent, and has the same relation to all the succeeding thoughts and actions which I call mine.

Reid thought that common sense required positing a simple, indivisible, and unchanging "I" that holds or guides the impermanent stream of conscious experiences. He insists that the self, the soul that makes me who I am, is simple or indivisible. Descartes said the same thing about our incorporeal minds. Yet Reid may be read as calling for some sort of clarification or elaboration of the Cartesian view. My incorporeal mind may be enough to explain the flow of experience, but it may not be enough to explain the "I" that stands in permanent relation to my experiences. We may have to show that a self or soul is, or needs to be assumed as, part of the Cartesian mind, or even add it as an ingredient beyond the mind, in order to gain a strong picture of personal identity and the possibility of personal immortality.

Indeed, this tactic is critical. Evidence abounds that all complex and divisible things are in flux and have at best relative permanency. So if the self is to be permanent and abiding, it had better not be complex and divisible. Still, one doesn't want to employ a premise simply because it is needed to gain the desired conclusion. We need to look closely at the evidence for the claim that the self that I am—that constitutes my essence—is simple and indivisible.

I Am Simply Indivisible

Let us suppose, therefore, that there is evidence for insisting on the simplicity and indivisibility of the self. What would it consist of? One possibility is that it is phenomenological or introspective. When I look for what constitutes my essence I am looking for something that does not change. Whether we find such a thing depends on our introspective acumen. But if we look with an open and unclouded mind and introspect a simple and indivisible self, this gives some credibility to the idea that there is such a thing. And indeed, on one straightforward reading, Descartes claimed to directly observe or see a simple, indivisible essence that is the soul. Furthermore, he was operating with the utmost intellectual diligence dedicated to doubting anything for which he had the least reason to doubt. While in this state of doubting whatever can be doubted, Descartes discovers first that he is a thinking thing, the famous *cogito, ergo sum.*[5] Even if everything else I believe is false, the belief that I am thinking cannot be false. That I am a thinking thing is self-confirming, since it is necessarily required for me to doubt, to wonder, to survey my beliefs. Later, with the "cogito" under his belt and a renewed confidence, thanks to two fallacious proofs for the existence of a good God who will not deceive him, Descartes says: "[W]hen I consider the mind, that is to

[5]St. Augustine—*si fallor sum* ("if I am doubting I am")—said it first.

say myself as I am only a thinking thing, I cannot distinguish any parts, but apprehend myself to be clearly one and entire."

One problem with this alleged apprehension is that it is undermined by Descartes's own view of psychology. The mind, according to Descartes, is possessed of reason, imagination, will, memory, perception, and the passions. Almost everyone will acknowledge that these mental faculties or capacities can be distinguished phenomenologically. That is, there is a detectable introspective difference when I imagine a state of affairs, when I perceive it, or when I will to create it. To which one might reply that these are functional distinctions of a single mind, but they do not name parts in the sense under discussion. That is, they are not based on underlying physical distinctions. They are not essentially related to anything my brain is doing but are, at most, analytically useful ways of demarcating the different things my indivisible mind can do.

But this reply begs the question. Surely the alleged fact that deploying my various distinct mental faculties does not essentially involve certain faculty-specific brain parts is not something anyone can introspect. Introspection cannot tell me what brain regions are active when I imagine, feel emotions, perceive, or engage in logical thought. Contemporary mind scientists distinguish between the surface structure of mind and its deep structure. According to current thinking, we are aware of our own mental states at a superficial level and in a coarse-grained way. I am aware that there is a tree in my path, but I have no conscious access to the neural structure of my perception. The tree's color, its shape, its distance from me, its separation from the background, the word "tree" and many other attributes are all processed separately and assembled into a single picture for presentation to my conscious awareness. But I have no access to any of this processing; I cannot, for instance, selectively examine or turn off the part that sees the tree's leaves as green.

No one thinks that a tree is only as it appears. It has deep structure—there are all sorts of facts about trees that are not revealed in ordinary perception. The same is true for perception itself. Every respectable mind scientist believes that perceptions involve complex brain processes not accessible to consciousness. Furthermore, we can explain why not revealing the neural texture of mind is a good evolutionary design strategy. We would be pathetic at survival if in addition to perceiving trees in our path we had to attend to the massively parallel unconscious processing that results in that perception.

One might defend Descartes along these lines. Descartes was a dualist. He offers at least three proofs for the absolute distinctness of the mind and the body. The mind is unextended thinking stuff (*res cogitans*) and the body is extended and unthinking (*res extensa*). The mind furthermore is transparent to its owner. When it comes to mental states what you see is what you get, and what you get is all there is. With the mind, there is no appearance–reality distinction the way there is for trees.

Descartes could have given this response because he, in fact, believed what has just been attributed to him. But we can't use this as a defense, since his arguments for mind–body dualism are all considered logical failures. Still, this defense, despite failing miserably, does help us see why Descartes's view of mind failed. His mistake was to take the fact that mental states do not seem physical—they don't seem to be hard or soft or to possess shape, color, or weight—as evidence that they are not physical. But in making this assumption Descartes sidestepped the reasonable suspicion that mental states might be like everything else in possessing deep, hidden structure, and thus being more than they appear.

Reid, like Descartes, seems to favor an introspective or phenomenological strategy to gain the conclusion that the self is simple and indivisible: "[I simply apprehend] that [my] self or I . . . is permanent, and has the same relation to all succeeding thoughts and feelings which I call mine." If the conclusion were based on introspection, Reid, like Descartes, would have a serious problem on his hands. Many smart and open-minded people will claim to have trouble seeing what he sees: a permanent, simple, and indivisible self. Hume, for one, famously wrote: "For my part, when I enter most intimately into what I call myself, I always stumble on some particular perception or other, of heat or of cold, light or shade, love or hatred, pain or pleasure. I can never catch myself at any time without a perception, and never can observe anything but the perception."

Phenomenological data can sometimes be decisive. When Descartes insists that everyone can perceive themselves to be a thinking, feeling thing, he expresses a truth that virtually everyone will assent to. The apprehension of a permanent, simple, indivisible self is not so universal. Hume couldn't see it. Nor, I claim, can you. The phenomenology of experience does not support the alleged observation.

This is why the best route to the conclusion that the soul is simple and indivisible is an inferential, not an introspective one. Let us try a kinder reading of Thomas Reid. Rather than see him as trying to win his prize on the shoulders of universal introspective evidence, we may read him as arguing that there must be an explanation for how I reidentify my self as the same self given that novel thoughts and feelings occur to me over time, and I perform novel actions. Read in this way, Reid can be seen as not simply repeating Descartes's idea that my mind seems in some sense indivisible, but as arguing that we must posit a transcendental ego or soul to explain how experience is possible, or at least how experiences may accrue to me as mine.

Hume, in the famous quote above, is almost always read as claiming that the self is unobservable and thus that there is no evidence for the existence of the self. But what he in fact asserts is that "I can never catch myself at any time without a perception." Hume cannot catch his self out on its own, all by itself, as it were, outside some relation to the flux of experience. He need not be denying the existence of the self, only saying that he never finds it not in the company of some perception or idea. If we interpret Hume this way, then Reid can argue that the best explanation for how Hume can engage in self-scrutiny, why he calls his self "myself," and how and why he invariably detects himself in relation to the contents of his consciousness, is that there is an abiding simple self, a permanent soul that owns, creates, and contains the contents of the stream of consciousness. Even if we can't see or catch this self out on its own, we can infer that the abiding sense of oneself as the same cannot be explained by the everchanging experiential flux, but rather must be explained by something unchanging that guides, remembers, accumulates, and contains the flux.

The first thing to notice is that this argument contains a new and questionable phenomenological premise, namely that I do reidentify myself as exactly the same self over time. Many will rightly balk at this. Indeed, I claim that most people do not identify and reidentify themselves as exactly the same over time. There is thus no need to posit a permanent "I" to explain what we are doing when we reidentify ourselves as exactly the same over time, since we do not do that. If this is right, then the most we can say is that self-location and self-identification involve trying to capture some relatively stable and permanent aspect of ourselves.

Suppose we accept that this is our actual or usual practice. Perhaps this is still enough to warrant the conclusion that the best explanation for the sense of personal identity, even deflated to consist only of personal continuity, requires the postulation of a simple and permanent "I." But it doesn't. Once we accept that we only have a sense of our self as continuous, but not as exactly the same, the conclusion is unwarranted.

Consider this case: Water is H_2O. If there is water, then there is H_2O, these being identical. We might say, therefore, that water is essentially H_2O. But despite seeming homogeneous and thus in that sense simple, water is in fact composite, it is composed of molecules. Now consider a cloud. Clouds exist and we reidentify a cloud as the same even as it moves across the sky and undergoes changes of shape and composition. Water droplets condense and evaporate many times a second. What is assumed in reidentifying a cloud as the same cloud over time? Do we need to infer something, "cloudness," that exists over and above the cloud that makes it the same cloud? The answer, of course, is "no."

Derek Parfit calls the view that there is a self or soul, or I, in addition to the life of my continuous embodied self, the "further fact" view. Clouds are dense collections of water and thus of H_2O. Do clouds contain water? Ordinary language will have us say "yes," which tempts us to think that something not the water must be containing the water. That is, that there is some "further fact" about a cloud than that it is just an airborne collection of water molecules. But there is no such further fact. A cloud is just a large, airborne, condensed collection of water. Furthermore, there is no ownership or containing relation between the cloud and the water droplets that comprise it. So long as it continues to exist, a cloud has relative—normally short-lived—permanency. But when it dumps its water in the form of rain, it is diminished and changed.

Brains are thought to be self-organizing dynamic systems. They function at different levels, from the microscopic level of single neurons, to populations of neurons or cell assemblies, to levels at which mental functions are experienced first-personally in perception, memory, and metacognition (which includes introspection). Higher and lower levels interact in both directions. But there is no single control center of the whole system, nor is there a single site that accumulates and houses the representations that the system works with and over. Thus there is no thing, other than the whole system, to which these representations belong. Self-organizing dynamic systems have a kind of permanency. They exist so long as their parts function in concert. William Calvin asks us to think of the mind at work as a "cerebral symphony."

This property of being a self-organizing dynamic system allows us to explain each level of mental functioning in terms of the whole system without ever going outside it. Just as an orchestra produces a symphony by the interaction of all its players, the brain produces thought and feeling by the interaction of its neurons, its neurochemicals, and its connections to the rest of the body and the environment.

Reid's argument requires going outside the system to explain it. But the existence of self-organizing dynamic systems establishes that we can best explain some systems by staying inside them, and it suggests that the brain, or better, the whole body in interaction with the environment, is the basis of all mental life, including the sense of self and, to whatever degree it exists, the sense of self-permanency.

Neither Descartes nor Reid was positioned to explain mental life in terms of self-organizing dynamic systems theory. But they were rightly worried about analogies like the one comparing a soul to a cloud. A cloud is complex, divisible, and impermanent. For them, the soul, however, is not like this. It is simple, indivisible, and permanent. Descartes and Reid both claim to find these features by introspection. But if I am right, few will go along with this claim, which means that the alleged simplicity and indivisibility of the soul are not based on credible introspective evidence. Indeed, the introspective evidence runs the other way, yielding a sense of one's self as at most relatively continuous and relatively permanent. If we only need to explain our sense of a continuous self, not of a permanent self, then the "I" that binds my experiences together does not need to be simple, indivisible, and permanent. If no sense of a permanent self needs to be explained, then no source of permanency needs to be invoked. Looked at in the light of actual introspection, the inference that an unchanging self, or soul, explains these facts is superfluous. It explains something that doesn't exist. Continuity of consciousness can easily do all the explaining necessary. My sense of myself as continuous, as the location of this self, this stream of consciousness, is easily explained by the fact that I am an organism possessed of conscious memory.

To sum up, the claim that my self is simple and indivisible either has introspective or phenomenological warrant or it doesn't. It doesn't. Still, the argument need not be bad or wrongheaded, so long as there are some facts, phenomenological or otherwise, that require as an inference to the best explanation that we posit a simple and indivisible soul. But there are no such facts in need of explanation. The only candidate is the alleged fact that, by virtue of being a person, perhaps because I have been created in God's image, I possess a simple, indivisible, immutable, and immortal soul. But this alleged fact assumes the truth of the premise in question. This is known in philosophical circles as begging the question.

The upshot is this: Most people believe they possess immutable souls, and that this soul constitutes their essence. But the arguments we have just examined—the best arguments ever produced—give no reason to think this. Perhaps there are other ways to gain the desired conclusion, but I think not. The very idea of the soul is in conflict with the way the scientific image conceives of things. But not, as we shall see, for all the reasons one might think.

Headless Horsemen

Oddly enough, no variation of the amputation thought experiment suggests decapitation. But insofar as the experiment is intended to provide ground for the idea that whatever I am is not constituted by my body as a whole or by any of its parts, the loss of the head should pose no more of an obstacle than arm or leg amputation. Of course, decapitation, unlike these other forms of amputation, will cause death. But again, that should not cause a problem. The argument aims to establish that my soul "has no need for any place; nor does it depend on any material thing; so that this 'me', that is to say, the soul by which I am what I am, is entirely distinct from my body . . . and even if body were not, the soul would not cease to be what it is."[6]

My essence is something simple, indivisible, and if Descartes, Reid, and many others are right, it is easy to imagine that my soul can exist without any body part, even my brain. From the perspective of the scientific image, by contrast, the fact that I am a thinking-feeling thing has everything to do with my possession of a brain that is housed in a body that interacts with a physical world.

Essences

What is an essence? The basic idea is that things possess both essential properties and contingent ones. Essential properties are attributes that are permanent and unchangeable, at least insofar as a thing continues to be what it is. In the standard view, my soul is my essence, my brain and body are contingent. In ordinary usage, the structure of my house is one of its essential features, its color is a contingent or accidental feature.

Perhaps the idea of the soul can be trashed, without even examining arguments such as Descartes's and Reid's, if science does not countenance essences. But science does countenance essences in the form of what are known as "natural kinds." Water is H_2O, salt is NaCl, and gold is the substance with atomic number 79. These identities tell us the essential attributes of water, salt, and gold, respectively. It follows that if the attempt to locate the essence of a person is loony, it is not because the scientific image tells us the search for essences is itself loony.

Another possibility is that science does not countenance unchangeable things. This tactic works better to gain leverage against the idea of the soul. Neither biology nor physics is friendly to the idea

[6]There is an exit strategy! I need my brain to interact with my body. But I don't need my brain to exist.

of unchangeable things. Every occupant of the natural world is in flux. Organisms are born and die, chemicals recombine, elements are transmuted within the bodies of stars, and the universe itself has a beginning, an evolutionary trajectory, and possibly at some distant time some sort of end.

The scientific image in our time might thus seem to side with Heraclitus over Plato. The world is all Becoming with no place for permanent, immutable Being. But this would be overstating things, since some well-respected physicists believe, just as the Greek atomists in the fifth century B.C.E. did, that the most fundamental particles or strings that make up the basic components of the natural world never go out of existence. These elementary particles, or strings, or whatever they turn out to be, participate in all manner of reconfiguration in the flux. But they themselves are immutable.

A soulophile might sensibly latch onto this last point and claim that she thinks of the soul as akin to fundamental particles. Souls partake of the flux, but each soul is permanent and immutable. Plato thought that each soul exists for all eternity, temporarily occupies a body, and then continues to live on for all eternity. And this, the argument might go, is perfectly analogous to the way some contemporary physicists think of the permanent and immutable basic constituents of the natural world.

To sum up so far: The image of an immutable soul will not conflict with the scientific image simply by virtue of its commitment to the existence of essences, since science allows for essences.[7] Nor does it conflict with the widespread scientific commitment to the universality of change or flux. One can, as we've just seen, believe in flux and yet consider some entities permanent and immutable.

---

[7]There is, however, a problem for the soulophile if he wants to use the sort of essences that I claim science allows to make room for the soul. The problem is this: I have said that science allows essences of the form "water is H_2O." Whenever there is a sample of water it is H_2O. But once the water evaporates both the water and the H_2O, these being the same, are gone. If an essential feature—suppose it is an incorporeal soul—that makes a person who she is, is discovered, but attaches to the whole organism or to some part, then we can speak of a person's essence—an essence constituted by her soul. But we will have been given no reason (yet) to suppose that it lasts beyond the organism's life.

The standard view in the West is that souls enter and exit the natural world by entering, visiting, interacting with, and then transcending or exiting the human body at death. Christian souls, unlike Platonic souls, come into being at a certain time, for example at conception, and thus are not eternal. According to standard Christian theology, souls do not exist for all eternity at the right hand of God, awaiting the moment at which he implants each soul-in-waiting into a body. Rather, God creates each soul and implants it into a body at the same time. Once a soul is created and implanted in a body, it never goes out of existence. Thus we can say that Christian souls are immortal but not eternal.

In any case, the key attribute of both Platonic and Christian souls is that they aren't physical and thus don't have physical properties. My body and its brain are impermanent and changeable, so these can't be what make my soul. Organic things die, decay, and disperse. A soul doesn't do so, so my soul is not organic. Nor, presumably, is my soul housed in one of those immutable elementary particles or strings that physics allows. These permanent and immutable elementary particles or strings—supposing there are such things—are the possibility proof that there can be permanent, immutable things. But they are not the right sort of thing to be or to house a soul. At least no one, to my knowledge, has tried to argue that souls are part of the universe as revealed by superstring theory.

The point at which the belief in souls will clash with the scientific image will be over the commitment to the soul's nonnatural, possibly supernatural properties. Contemporary science countenances all manner of weird and counterintuitive phenomena. But everything it countenances is part of the natural fabric of things and everything involved in the flux interacts according to the laws of electromagnetism, gravity, the strong and weak nuclear forces, and quantum field theory. But souls, in almost every view, are not part of the natural fabric of things and are not obedient to these fundamental laws. Perhaps they do not disobey or contravene these laws, but they operate in a sphere outside them.

The trouble is that the soul interacts with the body. Even if it is some third thing beyond mind and body, it has causal relations with the body via the mind to which it attaches, or which it contains, or whose its experiences it binds together. But according to this line of argument, it does so in undetectable ways. Occam's razor advises us to avoid postulating causes that are unnecessary to explain what needs to be explained, especially when the alleged causes do no work not done by principles or laws already known, and especially when their proponent is willing to concede that the work they do is necessarily undetectable.

One might think that this last feature of the scientific image—that it does not allow unnatural things or processes, and resists new and unnecessary fundamental laws—would, over time, have resulted in rendering the belief in a nonphysical soul a quaint idea, an understandable, even a compelling myth that humans once believed, but that contemporary folk have no use for. But as I said at the start, if you think this, you are mistaken.

The Root of the Illusion

The defender of the manifest image will claim that there are evidence and arguments to support the existence of immutable nonphysical substances. The defender of the scientific image will say that the evidence and arguments fail to support the desired conclusions. I have tried to show that the best arguments for an immortal soul do not succeed on their own merits. I have also tried to show that any argument that tries to yield the conclusion that humans possess nonphysical souls will butt heads with the scientific image.

One can imagine this response: "Okay, so the scientific image is not open to the possibility of souls, or of anything nonphysical. The manifest image is committed to the existence of souls. It's a matter of opinion, and I choose to go with the manifest image. The fact that science and philosophy do not countenance souls just shows that science and philosophy are close-minded. They haven't after all shown that souls are impossible."

To which this reply must be given. It is true that neither science nor philosophy has shown that souls are impossible. Furthermore, science is, or should be, open to explaining any fact or phenomenon that is in need of explanation. But the burden is on the soulophile to produce some fact, or evidence of some phenomenon, that is in need of explanation and that would make the inference to a nonphysical soul reasonable, if not irresistible. This has not happened. The scientific image need not be envisioned as being, in principle, closed to countenancing nonnatural entities or processes. It is just that no phenomenon ever observed has warranted invoking nonnatural entities or processes.

Revelation

The defender of the soul might turn to revelation as his evidentiary source. But doing so will likely cause him more rather than less trouble. The first and most familiar problem with claiming that revelation provides reliable evidence is that revelation is hearsay. The second problem is that the revelatory writings of different traditions conflict with each other: Their metaphysics and theologies differ, as do their proposed ethical codes—and they differ in substantive ways that make a difference (for example Jesus is God, Jesus is not God). The third problem is that even the revelations of a single tradition contain internal inconsistencies. Three out of the four New Testament gospels state that heaven is "out of this world," whereas one has Jesus saying it is here, in this world.[8] Which is it?

Furthermore, revelation contains odd commitments from which those who abide the tradition will rightly want to distance themselves. In the summer of 2000, after the TV guru Dr. Laura, "a born-again Jew," quoted scripture to condemn homosexuality, Dave McKee of Arizona State University circulated this response over the Internet:

> Dear Dr. Laura,
>
> Thank you for doing so much to educate people regarding God's law. I have learned a great deal from you, and I try to share

[8]In the summer of 1999, Pope John Paul II stated that heaven and hell were not actual places but rather that they were states (of this life) involving being in relation with God or out of relation with him. I was in Costa Rica when the pope said this, and the newspapers were filled for weeks with letters from disgruntled Catholics wondering if the aged pope was losing his mind. My friend Marcel Kinsbourne, a distinguished cognitive neuroscientist, told me that American and European Catholics, including members of his family, had similar reactions.

> that knowledge with as many people as I can. When someone tries to defend the homosexual lifestyle, for example, I simply remind him that Leviticus 18:22 clearly states it to be an abomination. End of debate. I do need some advice from you, however, regarding some of the specific laws and how to best follow them.
>
> When I burn a bull on the altar as a sacrifice, I know it creates a pleasing odor for the Lord (Lev. 1:9). The problem is my neighbors. They claim the odor is not pleasing to them. How should I deal with this?
>
> I would like to sell my daughter into slavery, as it suggests in Exodus 21:7. In this day and age, what do you think would be a fair price for her?
>
> I know that I am allowed no contact with a woman while she is in her period of menstrual uncleanliness (Lev. 15:19–24). The problem is, how do I tell? I have tried asking, but most women take offense.
>
> Lev. 25:44 states that I may buy slaves from the nations that are around us. A friend of mine claims that this applies to Mexicans but not Canadians. Can you clarify?
>
> I have a neighbor who insists on working on the Sabbath. Exodus 35:2 clearly states he should be put to death. Am I morally obligated to kill him myself?
>
> A friend of mine feels that even though eating shellfish is an abomination (Lev. 10:10), it is a lesser abomination than homosexuality. I don't agree. Can you settle this?
>
> Lev. 20:20 states that I may not approach the altar of God if I have a defect in my sight. I have to admit that I wear reading glasses. Does my vision have to be 20/20, or is there some wiggle room here?
>
> I know you have studied these things extensively, so I am confident you can help. Thank you again for reminding us that God's word is eternal.

I regret if in quoting this letter I am perceived as being glib or irreverent toward the Bible. But despite containing abundant historical, ethical, and spiritual wisdom, in some of the most beautiful prose and poetry ever written, the books of the Bible need much interpretation, and they can hardly be said to contain only wisdom about what is good, righteous, and beautiful. The person who believes that the Bible reveals God's word will need to explain why God has ideas that seem ungenerous, even weird. The usual response is to locate the source of oddities in the fallible composers of the biblical texts who did not hear God clearly. The trouble with this perfectly plausible argument is that we are provided no way to decide what portions of the Bible get God's word right and which don't.

In its present form, the New Testament contains gospels by Matthew, Mark, Luke, and John. These, as I have said, are not completely consistent. Imagine, for the sake of argument, that Jesus had written of his own life and conceived it in the way depicted by the Portuguese writer Jose Saramago in *The Gospel According to Jesus Christ.* In Saramago's novel, Jesus is a man of abundant goodness, born of the sexual partnership of Mary and Joseph, the lover of Mary Magdalene, who harbors no belief whatsoever in his own divinity. This vision is compelling, among other reasons, because Jesus is pretty much as Matthew, Mark, Luke, and John paint him, except he is not God incarnate. One might ask why this couldn't be the truth about Jesus. If the answer is that the weight of the extant gospels is enough to show that Jesus is God, then the question is straightforwardly begged.

That said, there is a burden on the defender of the scientific image to say why so many people are drawn to conclusions they think are unwarranted. There is a class of theories that seek to explain why people characteristically make some mistake in reasoning, which philosophers call "error theories." The Australian philosopher J. L. Mackie provided an error theory to explain the widespread, but in his view mistaken idea that moral badness is an objective property of certain actions. We say "murder is objectively wrong." But all we ever see are people who pull triggers and dead bodies. This is all that is objectively there. No one sees "wrongness." Mackie's diagnosis is that we project our powerful feelings about the wrongness of shooting innocent people onto the objective state of affairs and think mistakenly that the wrongness is in the state of affairs. But it is not. We judge, subjectively as it were, that murder is wrong, and there are of course good reasons to think so. The error is thinking that the wrong-making features of killing an innocent are out there, when actually they are in us.

In *Totem and Taboo* and *The Future of an Illusion,* Freud advanced an error theory to explain the belief in God. The belief in God is the result of a set of complex wishes, anxieties, and fears, including feelings of helplessness, the desire for safety, the wish to be protected, and the fear of a punitive father. Similarly, Ludwig Feuerbach, who influenced Karl Marx to think that religion was "the opiate of the masses," claimed that the image of God espoused by different civilizations was typically a transparent projection of an ideal of the kind of person most valued by that civilization. Spartans favored brave warrior gods, Jews forced into exodus favored a vengeful god, and Christians who disliked facing lions in the Roman Coliseum favored peaceful, loving coexistence and projected a kinder, gentler god.

The defender of the scientific image will need to provide an error theory to explain why so many people believe humans possess a permanent, immutable soul. This can be done. It might seem odd to suggest that the widespread beliefs in both God and in an immortal soul are made compelling by our biological natures, since this is tantamount to saying that the forces of natural selection have made us prone to falling into the grip of these illusions. It is paradoxical, to say the least, to suppose that our thoroughly biological animal nature makes us prone to invoke nonbiological spirits to explain ourselves and our world. But this is exactly what happens.

Let me explain, in a nutshell, why the posits of a soul and of a personlike spiritual creator are so appealing despite being unwarranted by the evidence. Four features of our cognitive-conative economy are the culprits: our natural disposition to acquire information inductively; our disposition to discern causal relations; our natural attachment to our own survival and that of certain conspecifics; and, finally, our capacity to hold time—or better, to hold events in time—in memory, to look from the present into the past and future. These four attributes explain why we are so prone to believe we possess an immutable soul.

We are wired to think inductively. By this I simply mean that members of the species *Homo sapiens* (and most other smart animals as well) are wired to detect patterns in nature by deploying the straight rule of induction. The straight rule involves reasoning (not necessarily consciously) that "if I observe that regularity R has occurred m/n of the time, then it will occur m/n in relevantly similar circumstances." Day has always followed night, therefore day will always follow night in the future.

An especially useful class of patterns involves causal regularities. That day follows night is a regularity but not a causal one, since neither night nor day causes the other. Rain, on the other hand, causes streams, ponds, lakes, and rivers to fill with water and it causes vegetation to become more lush and thus food to become more plentiful. The reason the discovery of causal regularities is especially useful is that it can lead to gains in control. Knowing that water causes plants to grow might lead me to divert a stream to create an irrigation ditch around a food source.

Our natural tendencies to try to discern regularities in nature and in the behavior of other people, and in particular to discern causal regularities, are fitness enhancing. These capacities are natural and original—they are innate; and they suit us with the right sort of equipment to survive long enough to reproduce. Every parent has noticed that children ask "Why?" a lot. So prevalent is this asking of why questions that many developmental psychologists have found it fruitful to think of the child as akin to a novice scientist. Our thirst to discover regularities, to discern cause and effect, to wonder and ask why, is a biological adaptation.

Next, add the fact that we are conscious creatures. We are aware of many of the things we observe. We are also aware of many of our own thoughts, our feelings and emotions, and of ourselves as some sort of continuous locus of thought and feeling. Consciousness, as William James argued, has a streamlike phenomenology: It is serial and accumulative. Past experiences are carried forward in the stream, some as conscious memories, and all contribute to the way the stream flows. They help constitute the surround, what James called the "penumbra," the shadow.

Regarding the emotions, we can say this: Thanks to a hypothesis proposed by Darwin in *The Expression of the Emotions in Man and Animals* (published in 1872), which was confirmed by Paul Ekman and his colleagues in the 1970s, it is safe to say that certain human emotions are universal. Fear, anger, surprise, happiness, sadness, disgust, and contempt surely make the list. Quite possibly embarrassment, shame, guilt, and remorse as well. There are several reasons for saying this. First, homologues of the emotions we experience and express appear in other animals—in canines, as well as in close ancestors. Second, in social mammalian species there are characteristic movements of the facial musculature that conspecifics recognize for the behavioral dispositions they display and then seem to respond to appropriately. Third, for the basic emotions of fear, anger, surprise, happiness, sadness, disgust, and contempt, we are locating better and better physiological and neurophysiological markers that distinguish among the different emotional expressions. Fourth, the facial expressions that signal these universal emotions are displayed and recognized across all human societies—among preliterate Papua New Guinean Highlanders as well as native New Yorkers.

The basic emotions are part of the original equipment with which we enter the world, just like our dispositions to make inductive inferences and to seek causal regularities. As with these other dispositions, we are naturally disposed to feel and express the basic emotions. Taken together, these emotions constitute exactly the sort of conative economy one might expect in a creature that cares about its own survival, and the survival of certain others. The basic emotions are part of the original and natural equipment of *Homo sapiens* and they come into play when fitness is at stake. When my physical well-being or survival (or that of others for whom I care) is threatened, fear and anger are triggered. When eating or mating opportunities arise, I am happy. When I lose a loved one, I am sad. It is plausible to think of the basic emotions as fitness enhancing, as adaptations.

So far we have a picture of humans as naturally possessed of cognitive dispositions to inductively locate regularities in nature, including causal regularities, and as conatively disposed to want to survive and thus to express feelings in certain universally recognized ways. Furthermore, individual persons experience their own consciousness as a serial, accumulative stream of experiences.

The final ingredient needed to explain the illusion of an immutable soul involves adding into the mix the powerful desire to survive. The powerful attachment to continued existence, the powerful desire to want, at every juncture, to survive, and thus to make efforts always to survive, makes perfect evolutionary sense. We are conscious beings-in-time, and we are able to detect regularities about the external world, the social world, and our own histories, and to hold these in conscious memory. Among the regularities we detect is that humans, that is, all creatures like us in all relevant respects, are born and die. Early humans would have been vividly aware of the messy, smelly, maggot-infested way in which dead bodies decompose. But they would not have needed this awareness to have an instinctive fear of death.

Here is an account of how the belief in a soul able to exist after bodily death finds fertile soil.

- *Consciousness Is Not Physical*: Physical things are created and destroyed. Conscious desires, feelings, emotions, and thoughts do not seem like ordinary physical things. They do not seem physical at all. A rock is solid, it might be too heavy to lift, be oblong, and reddish. But thoughts and feelings are not solid and do not have weight, shape, or color.
- *The Possibility of Permanence*: Even if physical things are impermanent, it is possible that nonphysical things are permanent.
- *Mental Causation*: My conscious thoughts and desires cause certain things to happen: I want the crops to grow so I irrigate them. This is the possibility proof that nonphysical things (my conscious thoughts, feelings, and intentions) can cause physical things to happen.
- *The Phenomenology of "I"*: My stream of consciousness is not only mine, it moves forward in time. "I" accompany the ride upstream. It is "my" accompaniment that makes this stream "my stream." I have strong feelings that my self—that "I"—will continue to exist even after my body dies. I already have evidence that "I" survive all manner of bodily change. Why not death?
- *Annihilation Is Not an Appealing Idea*: I am powerfully attached to the hope that I (and my loved ones) do not stop existing

> when my (their) body(ies) does (do). I fear my own death and am saddened when loved ones die.

All things considered, the belief in a nonphysical soul that constitutes who I am is a reasonable inference based on the observation of past regularities (my continuous existence); the nature of my consciousness (it does not feel physical); and the lack of evidence that nonphysical things are subject to the same principles of impermanency and annihilation as physical things are. Furthermore, the postulation of a permanent soul that accompanies my earthly life is supported by phenomenological evidence—I experience myself as permanent and as, in some strong sense, abiding.

The response to this line of reasoning—or better to the amalgamation of this set of beliefs—is not to say that it is silly but that it overreaches. First, we have already seen why the fact that consciousness does not *seem* physical, and that mental causation *seems* to involve nonphysical desires or intentions causing such things as talking a walk, should not be taken as evidence that these things are as they seem. The evidence points to a physical basis for both mind and for mental causation.

With respect to the alleged phenomenological fact that a permanent "I" accompanies me through life, the situation is even worse. For most people who honestly and open-mindedly examine the stream of conscious experience, it doesn't even *seem* as if there is a permanent "I." Second, and even more importantly, it is well known that we need to be watchful of beliefs that are rooted in hopes. Wishes and hopes carry no epistemic weight in and of themselves, but they do create powerful impulses to believe that what one hopes will become true. We can even say that one of the natural functions of wishing or hoping is to make the individual who wishes or hopes more capable of bringing about the state of affairs she hopes for. But wishes and hopes can work, of course, only in cases where the agent has the power to make the hoped-for state of affairs happen. Normally we make wished-for states of affairs true by doing things. I wish to eat, so I go to the refrigerator. But not even the soulophile's most powerful hope is thought capable of making the existence of souls more likely to be, or to become, true. Souls are permanent or impermanent independently of any wish or hope. The hope is merely a powerful but inert desire. Therefore, if we are to save the idea that we possess a soul, a permanent and immutable self, we will need to look at other grounds for the belief.

One might imagine this ill-advised attempt to provide such a reason. Okay, Flanagan, you say the belief in an immortal soul is made compelling by the way Mother Nature designed us. Fitness accrues from true beliefs, so surely if Mother Nature pushes us powerfully and relentlessly to believe something, it must be true.

This argument won't work. Evolutionary biologists are quick to point out that many biological traits are satisfactory but not optimal. Mother Nature is a tinkerer who builds on designs that work well enough but that are seldom the best that could be designed if there were no prior constraints. Expanding on an example from Steven Jay Gould, the philosopher Karen Neander writes:

> The Panda's "thumb" was selected for stripping leaves off bamboo and that therefore is its function. But it does not follow that the Panda has the best bamboo-leaf stripper it could possibly have, or that natural selection worked toward an adaptive outcome in this case in the absence of all constraints. Far from it. While the Panda's thumb has a clear function, it is a wonderful illustration of the fact that natural selection is a tinkerer and a satisficer, heavily constrained by the past and by the alternatives that are presently available. The Panda's thumb—in fact, an elongated wrist bone—is an imperfect design from a design engineering point of view.

Our inductive and causal reasoning capacities are excellent examples of satisfactory but nonoptimal designs. Humans, for example, are notoriously prone to fallacies of the small sample, and what is different, fallacies of unrepresentative samples. When it comes to making causal inferences, small sample size and unrepresentativeness trip us up again and again. Furthermore, we are constantly inferring causation when mere correlation obtains, for example, since night always follow day one might infer that the link is causal. Which, of course, is false. Or better: Most people think cold, wet weather causes colds—whereas a virus that is more abundant in winter does.

Aristotle spoke of "the total cause" of an event, meaning the specification of everything that contributes to making the event happen as it does. He was aware that we are almost never positioned, nor do we normally aspire, to give total-cause explanations. For pragmatic reasons, we normally restrict our explanations to factors that are in close temporal proximity to the event to be explained, and that are still under our control. Thus if you ask me why my crops are doing so well, I will explain that I water them and water causes them to flourish. You might ask why water does this, where the water comes from, and so on. If you ask me how that plant got here in the first place, I may explain that I planted a seed. You might ask where the seed came from, which I might explain as a product of another plant. To which you might ask me when, where, and how this whole seed to plant cycle got going, and so on.

Our untutored reasoning capacities do best when operating at the human scale, dealing with causes and effects readily accessible to our ordinary senses (for example, "water" as opposed to "H_2O"), and covering relatively short spans of time. We are fairly good at providing local, incomplete, and coarse-grained causal explanations, but not at providing total, fine-grained causal explanations. Ordinary humans are fairly good at detecting regularities at a coarse level, but not at fine-grained levels (this being science's job).

But, and this is the rub, humans are very prone to overreaching. The little child may eventually stop asking "Why?" at every turn. But that little child remains in each of us—silenced perhaps, but with the wheels still churning inside. One of the most widely heralded set of findings in cognitive science in the 1970s showed how easy it is to get smart adults to make consistent errors in inductive reasoning, to confabulate utterly implausible explanations of their own behavior, and to confidently claim to know how their minds operate at fine-grained levels to which they have no possible access, for example, to explain confidently how unconscious language-processing works.

The evidence is that our inductive capacities and cause and effect reasoning abilities work satisfactorily when dealing with the sorts of problems humans likely faced throughout the Pleistocene. But they did not evolve a strong sense of their own limits. Prod me, even gently, to say more than I know and I will produce some noise, but not, in all likelihood, knowledge.

Regarding the claim that we each possess a soul that is simple and indivisible, that is nonphysical, and that is immortal, one lesson to take home is that each claim involves saying much more than we can possibly claim to know. Our excessive confidence in these beliefs requires explaining. I claim that the explanation lies in the strength of our wish to be, to survive, for the lights of consciousness never to burn out. But this wish provides absolutely no evidence for its own truth. Furthermore, it is easy to explain why, once the belief takes hold, it would spread like wildfire. It is appealing after all.

One final point. I've just been responding to the argument that the belief in a permanent self as well as an immortal soul is made compelling by the way Mother Nature designed us, and that since fitness accrues from true beliefs, these beliefs must be true. I have explained why this response fails. But I want to correct one possible misinterpretation. I never said that the belief in a permanent, abiding, immutable self was innate, nor that the related belief in personal immortality was. I don't think any belief is innate. All I claimed was that there was a natural way to explain how our kind of animal might come to have the thought that it was possessed of a nonanimal part—the self or a soul.

Ex Nihilo Nihil Fit

I have been making every effort to give the soulophile his best shot at making sense of and defending his claim that there is a permanent thing rightly called "the self," and that the self is best understood as "my soul." I dismissed revelation as a possible source of evidence for this belief since despite abundant references to souls, revelation is hearsay, and different revelatory sources make mutually and internally inconsistent claims. It is unclear, for example, how the claim "Ashes to ashes, dust to dust" sits comfortably with claims that we have immortal souls. Furthermore, some of the wisdom contained in the scriptures, as we saw from Dave McKee's letter to Dr. Laura, will be seen as rubbish by almost everyone's lights.

But I am intent on giving the defender of the soul every chance. One route to the desired conclusion that I have not explored is a theological one. Keep revelation to the side, but consider this possibility: Perhaps there are grounds for believing in God that do not depend on revelation and mindless cultural contagion, but rather on what is sometimes called "natural theology." Perhaps there are natural theological arguments that provide a compelling but indirect route to the conclusion that the soul exists. So far I have only considered arguments that go directly for the conclusion that humans possess souls. Perhaps there is another route. Prove that there is a God, and then show that this true belief warrants believing that we have souls. Here goes.

Aristotle said that all philosophy begins in wonder, in the sense of awe at nature's complexity and magnificence, and wondering, in the sense of wondering why. Why is there something rather than nothing? Why is there anything at all? And why am I one of the things that is?

Ex nihilo nihil fit—from nothing, nothing is made—is an axiom shared by ancient, medieval, modern, and contemporary thinkers, and it is assumed by the defenders of the scientific image (with quantum physics giving considerable pause with respect to the axiom's generality).

Everything that there is, was caused to come into being, and caused to be the way it is, by something else. The cosmos and all the things in it exist. Each thing, according to the axiom, was caused to be, and to be as it is, by virtue of something else. But this process can't go back infinitely, otherwise we'd have an infinite regress of causes and effects, movers and moved, changers and changed. There must have been a First Cause, a Prime Mover Itself Unmoved. This First Cause must be self-sufficient in the sense that it could not itself have been caused to come into being. It must have always existed. The Prime Mover always was. This argument is called the cosmological argument because it purports to explain how the cosmos came to be. We—many of us, at any rate—call the Prime Mover "God," the creator of heaven and earth.

Arguments for a permanent and immutable soul, as we saw, get off the ground—at least in large part—by combining the inductive recognition that I persist through time with the powerful hope that I will continue to do so. The cosmological argument takes certain inductive evidence, certain observations about the ubiquity of cause and effect relations, and runs them in the other direction—backward into the distant past, rather than forward into the future.

Besides this, there is another difference worth noting. We worried that the powerful desire to keep living, the hope not to suffer a painful and messy death and not to be annihilated as the person one is, provided suspiciously strong desire to believe what in fact is not in the cards—survival after death. The cosmological argument is cognitively purer, less open to the concern that the conclusion it yields is a matter of wishful thinking. The argument is based pretty exclusively on logic, evidence, and reason. Even one who finds the argument unconvincing will have a hard time pointing directly to some deeply cherished hope or wish that the argument conveniently supports.[9]

That said, belief in an immortal soul and belief in God are, as it were, a match made in heaven. In practice they are typically used to shore each other up and to elaborate each other. Humans are made in God's image. God provides a home for our souls after we die; both are nonphysical, and so on. My only point for now is that the two arguments are at first pass independent, and the cosmological argument does not engender the worry that it simply expresses a wish in reason's guise.

The cosmological argument, as I've expressed it, actually says nothing about the properties of the Unmoved Mover except that it is itself uncaused and that it is the cause of the universe. But two

[9]Freud, for reasons discussed earlier, thought that God's existence is based on wishful rather than rational thinking. And E. O. Wilson, the father of sociobiology, now dubbed evolutionary psychology in certain circles, sees a complex set of basic needs—for communal coordination and ethical control, the desire to survive—as being behind the origin of religious belief. Indeed, Wilson believes that genes that make people easy prey for religious conviction may have been selected for insofar as adherents might be more cooperative and altruistic than conspecifics that have more trouble learning the rules and falling under the spell of the relevant rituals and beliefs. Thus Wilson writes: "The predisposition to religious belief is the most complex and powerful force in the human mind and in all probability an ineradicable part of human nature." If Wilson intends to say that belief in God is innate, he will have trouble explaining, for example, why as many as 15–20 percent of contemporary Western Europeans are nonbelievers.

thoughts about the Unmoved Mover actually arise. It must be extremely intelligent and personlike, and it must be nonphysical. The first thought comes from certain intuitions about the design of the universe. These intuitions result in "it" becoming "he." Complex and intricate things—tools, houses, watches, and the like—are all the result of finite intelligent design. In the normal course of things, these artifacts do not pop up without the effort of intelligent designers, nor do they suddenly appear when nature takes unusual and unwelcomed turns. Hurricanes, earthquakes, fires, typhoons, and tornadoes do not leave in their wake new tools, houses, and watches. If these things were there before the natural calamity, we frequently find them destroyed. The universe is a magnificent and mind-boggling engineering achievement, surpassing anything that run-of-the-mill human minds could produce. Design takes intelligence requisite to the complexity and magnificence of the object. Thus the supermagnificent, supercomplex cosmos must be due to a superintelligent, personlike creator.

The reason the superintelligent First Mover must be nonphysical follows from the observation that everything physical is part of the flux and thus subject to change. That which causes there to be something rather than nothing cannot be part of the flux, otherwise he would be subject to the laws of cause and effect, change, impermanency, and dissolution. The Prime Mover must transcend all flux. He must be unfluxed, indeed unfluxable. Everything physical is part of the flux. Therefore, the Prime Mover must be nonphysical.

I said earlier, based on several not very subtle or sensitive polls, that somewhere between 70 and 96 percent of Americans believe in a personal God, one who is personlike except that he embodies every perfection. It is not merely that God is possessed of great intelligence; he is omniscient, omnipotent, omni-this and omni-that, where the "this's" and "that's" refer only to good and noble traits. A personal God, as normally conceived, is not only personlike, albeit infinitely or perfectly so, possessing finite humanlike traits in perfect form, but he also takes personal interest in humans. He is willing to communicate with us personally, through prayer, for example (recall that 40 percent of scientists believe God does so). If you watch professional sports you might come to think that God is so personally involved in human affairs that he roots for certain teams and assists particular athletes. God is, according to many athletes, The One who guides the winning shot, goal, or touchdown.

An Interlude: Gus the Baptist

The idea that the one true God works for his people, for the anointed ones, is widespread. God loves and cares for his people, but often not for everyone, especially not for nonbelievers. Perhaps athletes who think God aided them in making the winning shot think their faith was stronger than that of the athletes on the other team. I don't know. But the idea is certainly available across many traditions that belief in their God, participation in their rites, will alone be rewarded. Nonbelievers may well be left out in the cold, or dipped for all eternity in hot oil.

You all know about John, the Baptist, who anointed those who chose to follow Christ. But you are undoubtedly unaware of my Grandmother Augusta (Flanagan's) baptismal practices. In the late 1970s, when I was a graduate student living in Cambridge, Massachusetts, I visited Grandma Gus each Tuesday evening for supper. The rest of my family lived in New York, and Taunton, Massachusetts, her hometown, was only a 45-minute drive for me. Grandma had worked as a nurse for many years at Morton Hospital before retiring. Now in her mid-seventies, she was in failing health, but after a nice roast and a glass of whiskey, we would take a walk if the weather permitted. One spring evening, as we walked through town, a well-dressed middle-aged man greeted Grandma. "Mrs. Flanagan, great to see you! How is everything?" Grandma introduced me to the gentleman, who turned out to be the mayor of Taunton. After a few minutes of pleasant chitchat, we proceeded on our way. I remarked that the mayor seemed like a nice man and that I was pleased that he showed such personal concern for his citizens. Grandma replied that she had "known him forever. In fact, I baptized him." I knew (what you perhaps don't) that Catholic nuns, doctors, and nurses are all allowed to baptize a Catholic infant if a priest is not available and the child is in imminent danger of dying. The long version of the Catholic "Credo" (literally "I believe") includes these words:

> We believe in one Baptism instituted by our Lord Jesus Christ for the remission of sins. Baptism should be administered even to little children who have not yet been able to be guilty of any personal sin, in order that, though born deprived of supernatural grace, they may be reborn "of water and the Holy Spirit" to the divine life in Christ Jesus.

I was puzzled, and I told Grandma that the mayor's name sounded distinctly Jewish. To which she replied: "Oh yes, he is Jewish. I baptized all the Jewish children at the hospital." I was speechless and appalled. But I understood that for everything Grandma had been taught, and for everything I had been taught as a boy, not being born Roman Catholic constituted an emergency. In her own sweet, loving way, within the constraints of the queer exclusionary orthodoxy that Grandma had been trained in, she had liberally interpreted her privilege, as a Catholic nurse, to baptize in emergencies.

Is the Cosmological Argument Any Good?

Since only 9 percent of Americans at most think that scientific cosmology and the theory of natural selection can, in principle, explain how the cosmos came to be and how life came to exist without invoking a creator, it follows that at least 91 percent of Americans are at least sympathetic to the idea of a personal God or a creator (of unspecified attributes) who is the Prime Mover. Since the number of people who say they believe in a personal, creator God varies between 70 and 90 percent, it seems safe to assume that perhaps 20–25 percent of people are somewhat hesitant in their belief. This group doesn't think naturalistic explanation suffices (unlike the 9 percent who do), but their idea of what sort of nonnatural, theistic cause is necessary to explain the world is not firmly held. We might imagine that true believers would call many of these folks "agnostics."

Philosophers are in the business of asking for reasons for beliefs. Suppose the cosmological argument is offered, as it often is, as the reason for the belief in a God. Is it any good? That is, does the cosmological argument work? The answer is "no." The cosmological argument for the Prime Mover; the argument from design that gains the Prime Mover personlike qualities, albeit majestic, perfect "God-like" ones; and the argument for God's incorporeality based on the premise that only nonphysical things possess the sort of self-sufficiency that the Prime Mover must possess, are all bad arguments. If you or your friends believe in God because of these arguments, you believe in conclusions that logic will not support and in arguments that no decent logician will accept.

I must admit to an urge to stop writing. The urge is not caused by fear of offending people, although this is what I am about to do. The reason is this: I've explained why the arguments don't work to many smart students over the years. They almost always see why the arguments are abject failures. But many of them don't seem to care. I sometimes suspect that people just don't really care if some of their most cherished beliefs are rationally groundless. This is sad, if true. Or bad, since we should want our beliefs to be true.

Another possibility is that what many people gain from religious belief, and what they legitimately care about, is the ritual, the community, the shared ethical values, the coffee and donuts. The expressed theological commitments are not really intended to be statements of true beliefs, despite wearing such clothes, but rather they are mood-setters that, if said in a certain properly serious way, serve to get the right communal juices flowing. I actually think this is a stronger reason than many people will admit. But the fact remains that once the mood is set, people do say that they are committed to certain beliefs, and many a war has been waged over theological matters. Even if the number of people who genuinely believe in a God or an incorporeal and immortal soul is smaller than the polls report, some significant number of people do believe that they are speaking the truth and that they have evidence and arguments for these things. In fact they don't. Truth, as I have said, is not the only thing of value, but it is what I am concerned with. It is important that we realize when we hold beliefs that are warranted, as opposed to beliefs that have no warrant.

So what about the arguments for God's existence? We can make quick work of them. Bertrand Russell thought that the cosmological argument was laughable because he thought it flagrantly inconsistent. The argument insists first that *everything* that exists has a cause and then denies that *everything* that exists has a cause on grounds that this would require believing in an infinite regress, that is, cause and effect relations running backward for all eternity. Something must be uncaused, namely the Prime Mover, the First Cause. According to Russell's reading, the argument is committed to the assertion that everything is subject to the principle of cause and effect *and* to the denial that everything is subject to the principle of cause and effect. This is inconsistent, the worst kind of logical error since inconsistencies are incoherent, they are senseless.

One way to avoid this very bad result is to read the argument more charitably, as saying that everything we observe is subject to the laws of cause and effect but that inferring that all things—including the things we are in no position to observe—are so governed commits us to the belief in an infinite regress. An infinite regress is itself nonsensical, absurd. There is a way to avoid it: Posit a First Cause that always was, that was not itself caused to come into existence.

So far so good. The inconsistency that made Russell laugh is avoided since the argument no longer claims both that everything is caused and that not everything is caused. The new version claims that everything we observe is caused, but something that we do not observe must be uncaused. We are required to infer the existence of the Unmoved Prime Mover unless we are willing to accept the crazy, senseless idea of an infinite regress. We need not worry about what causes the First Mover to be as it is, because the point of inferring the First Mover is to put a stop to precisely that question.

But a new problem arises. The Prime Mover stops the alleged absurdity of an infinite regress by being attributed the property of always having been, of existing for all eternity. But in virtue of having existed as an Unmoved Mover for all eternity, the Unmoved Mover does not help us to avoid an infinite regress. It—or he—embodies an infinite regress! The First Cause is uncaused but infinitely regressive. The alleged absurdity of an infinite regress is not avoided, but rather embraced, by the cosmological argument. If an infinite regress is an absurdity and to be denied, the argument concludes by affirming the existence of something absurd. Any way you look at it the argument is a miserable failure.

If you appreciate this problem you might be led to the conclusion that the argument should not assume the absurdity of an infinite regress but should accept it—accept that there must always have been something rather than nothing.

But if the spirit of the argument shifts to embrace an infinite regress, we seem to have three main choices: infer a spiritual first cause (or regress of spiritual causes); assume a physical first cause (imagine, for example, that the singularity that imploded at the Big Bang always was); or assume an infinite regress of physical causes and effects that, for example, at a certain point caused the singularity that eventually banged.

This new, regress-friendly argument gives us absolutely no reason to prefer the posit of a nonphysical spiritual Being that always was over the physical alternatives. Thus it fails completely as an argument for God's existence.

I represented the argument from design as a way of giving substance to the Prime Mover, as a way of seeing our way clear to assigning to it personlike characteristics, especially intelligence—of making "it," "him." But this has often been used as an independent argument for the existence of a God who created heaven and earth. In either case, the idea, as we saw, is that majestic and complex designs are due to majestic sources. God, being majestic and incredibly powerful, is the source of the awe-inspiring universe.

Whether it is viewed as an addendum to the cosmological argument, as a way of filling out our picture of the nature of the Prime Mover (he specializes in universe design), or as establishing his existence on its own merits, this argument also fails.

The argument operates as an inference from an analogy with the human design of artifacts. Watches, clocks, dams, houses, pyramids, boats, ships, houses, skyscrapers, and so on are products of intelligent human design. These things never pop up without intelligent human design. The universe is unimaginably more complex and magnificent than any and all of these things. A God of suprahuman intelligence must therefore have created it.

This argument turns on a true observation about the capacities of human designers and their remarkable productions. Furthermore, it legitimately forces us to reflect on the awe-inspiring qualities and incredible intricacy of the alleged creation. But the argument does not work, for reasons made clearest by Darwin. According to Darwin, nature works totally by what Daniel Dennett calls "cranes," never by "skyhooks." A crane, which is itself grounded (and not all that complex), lifts up what is already there, sometimes making something stable and new. Skyhooks, if there were any, hang up there on their own as it were, drawing into being that which did not previously exist. God is a skyhook, natural selection is a crane. Darwin's theory yields a credible and well-confirmed alternative to the Genesis story of creation. Life as we know it not only could have but did evolve, without the work of an Intelligent Designer.

Many still find this idea incredible. They hold to their breasts the powerful intuition that the cosmos can't consist, or be the result, of a craning operation all the way down, all the way back in time. Some, seeking comfort in conciliation, will claim that God did create the universe, just not in the way Genesis depicts. He created a world in which there was a singularity that banged and that then proceeded eventually to give rise to life, which then unfolded according to the principles of natural selection. God, being omnipotent, created the singularity that banged, he created the laws of physics and natural selection, and by virtue of being omniscient he saw everything unfolding as it, in fact, has done.

Now, you can say this. Indeed, you can say anything you want. It's a free country and all that. But there is no evidence for positing this sort of God, or any other sort. The reason is this: Granted that the question of why is there something rather than nothing is gnawing, the cosmological argument offers no basis to choose a spiritual creator that always was over a world in which matter and energy always were. One might even plausibly claim that the posit of a nonphysical source of creation is worse than the naturalistic alternative, because we have no evidence that nonphysical things exist, whereas we have bountiful evidence that physical things exist.

Putting this thought to one side, there still might be some reason to choose an eternal spiritual creator over an eternal material world if, in addition to our observations about the grand productions of human persons, we had reason to believe that magnificent design could not occur without a Magnificent Designer. But Darwinism is the possibility proof that magnificent design can occur, indeed has occurred, without any involvement of such a Designer.

Still the theist might find some room to maneuver. She can claim that there is no inconsistency in positing a God as creating the resources necessary for the Big Bang to have occurred and for biological evolution to work. Indeed, she might claim that the idea of God as creator of the world that astrophysics and natural selection then proceed to explain in their own terms satisfies everybody. Many scientists would agree. No heads butt, no one steps on anyone else's feet.

The proponent of this view can legitimately point out that astrophysics does not try to explain the universe before the Big Bang, and Darwinian explanation only kicks in long after the Big Bang, when self-replicating molecules emerge. So not only is theism compatible with astrophysics and natural selection, it provides a pleasing answer to that pesky question—"Why is there something rather than nothing"—that does not tread on terrain that astrophysics and evolutionary biology claim to cover.

This sort of theism does appear to live comfortably with astrophysics and evolutionary biology. But I must still ask the philosopher's question: Even supposing this answer is pleasing or comforting, is there any reason to think it true? The answer is "no." The cosmological argument responds to a deep urge to answer the foregoing question, but it leaves us clueless about how to answer this question. Furthermore, astrophysicists and philosophers of science will, if they are honest, admit that they have no resources with which to speak of what happened before the Big Bang. They can tell you what the universe was like a few femtoseconds after the Big Bang. But about what it was like before it, they have nothing to say. Their reason for not speaking, however, is principled. Space and time began with the Big Bang. The very phrase "before the Big Bang" presupposes that we can sensibly think of time "before" the Big Bang, but we can't. Wittgenstein famously said "whereof one cannot speak one should be silent." Quietism about the matter is warranted.

But our conciliatory theist will claim that she has an unquenchable urge to speak about what was "before" the Big Bang, and thus to make sense of the terrain that the astrophysicists and evolutionary biologists can then help us navigate. But here she faces a problem, actually two problems. First, if the above objections to the cosmological argument are correct, then she has no reason to prefer a theological to a naturalistic infinite regress as a way of satisfying her urge. Furthermore, what she has been seeking can now be seen as an answer to a nonsensical question. Talk of time, of "before," has no place except in reference to after the Big Bang occurred. We don't know how to speak about the matter that concerns us without using the language of time and space. But astrophysics tells us that these concepts make no sense where we are seeking to apply them.

The upshot is this. The conciliatory position that claims no inconsistency between the belief in God who created the singularity that banged and the laws of natural selection that came into play when life was formed can provide no reason to posit a theological cause of the Big Bang over a fully naturalistic cause. More disturbingly, saying anything about the matter engenders inconsistency, for the astrophysicist will claim that talk about what was "before" the singularity makes no sense, because there was no time until after the Big Bang. The theist faces a dilemma: Either she posits a theological rather than a naturalistic explanation of the universe for no other reason than that she has a strong desire to do so; or she posits a theological source of creation that requires having thoughts about "before" that make no sense and are inconsistent with what our best science says is thinkable about time.

Recall that the reason for discussing the issue of God's existence, given that all direct arguments for the soul are abject failures, was to see whether the existence of a spiritual creator might be used to shore up the arguments that aim to establish that humans possess an immutable and nonphysical soul. If there was reason to believe in an immutable spirit who is Creator, then there exists at least one soul, one permanent, abiding, eternal, immutable spiritual Being. And if we could then provide evidence or argument for the idea that this spirit created humans in his image, the idea that we possess the equipment we have been seeking would gain plausibility.

But the discussion of God has led to the conclusion that there is no reason to believe in him. So belief in God, being unwarranted, cannot help the defender of the view that I am possessed of a soul, a Cartesian ego.

These two beliefs, in an immutable and incorporeal soul and in an incorporeal God, call for somewhat different attitudes. The failure of arguments that there is a soul warrants the conclusion that there are no souls, no Cartesian egos. The failure of the arguments that there is a God warrants quietism. We can say that souls don't exist because we have most, if not all, of the evidence for or against souls in front of us. Persons exist and they either do or do not possess souls. When we discuss the question of whether or not we have souls, we are dealing with the properties of things in the world we occupy. When we try to speak of God, we attempt to go outside the world, to transcend the concepts that are required for us to have coherent thoughts.

One might read what I have just said as a recommendation for agnosticism. But the quietism I am recommending is not quite like ordinary agnosticism, in which one expresses uncertainty about the existence of a God having the properties assigned to him by a particular religious tradition. The quietism I am recommending involves resisting the question, saying that I cannot speak on the issue (nor can you, despite your willingness to do so) because we have reached the limits of language. I am being asked to speak about matters on which I cannot coherently or sensibly entertain thoughts. If a quietist of the sort I recommend is coerced into saying something—suppose she is given a list of all the conceptions of God ever entertained by the world's religions, and required to say which one of them, including none of them, seems like the most likely suspect to exist—then I think she should say none of them. This is because the quietist thinks nothing sensible can be said about the matter. Her answer "none of the above" means that for any determinate conception of God, she will see both affirmation and rejection of that conception as making no sense. She is not saying "none of the above" to suggest that there is a true answer not on the list. The quietest thinks that there is nothing worth saying, nothing sensible to be said, either about any conceivable positive characterization of God or about the denial of any characterization. Some people will see this quietist as tantamount to an atheist, and that may be a reasonable way to understand her. But she is not an atheist who disbelieves a certain conception of God. She sees no basis to coherently believe to be true or false any claim for any God.[10]

Buddhism and the Scientific Image

The view I have promoted so far is similar to one held by most Buddhists. Of course, I did not approach either the problem of the self or the problem of the nature and existence of God from the perspective of Buddhism. The latter problem was destined to arise in our discussion of the idea that we each possess a simple, indivisible, incorporeal, and immutable soul. I am, in fact, a Buddhist practitioner, but the gist of everything I have said was worked out in my head when I was a teenager, long before I knew anything about Buddhism. Here I have approached these problems as a Western philosopher of mind and ethics—the current incarnation of that teenager who saw the problems with the beliefs in a soul and in God—interested in unearthing and working through the conflicts I see between the manifest or humanistic image of persons and the scientific image.

But it is worth pausing to reflect on the apparent accident that Buddhism, almost alone among the great ethical and metaphysical traditions, holds to a picture of persons that is uniquely suited to the way science says we ought to see our selves and our place in the world.

I deliberately called Buddhism a great ethical and metaphysical tradition, but not a religious tradition. Buddhism not only lacks a theology but actively rejects the theological impulse. And it does so for very much the same reasons that led me to recommend quietism on the nature of God. Humans lack the kinds of minds required to think

[10]One could be an impatient quietist. I fit the bill. When a person puts forward a determinate conception of God, for example, "He is possessed of every perfection," one can try to show why this concept is unstable in its own terms. For example, the problem of evil is typically brought on the scene to show that if God is said to be possessed of every perfection then he is powerful enough to abolish all evil and suffering, and if he is all-good and all-loving, he should want to abolish evil. But evil is abundant, so God doesn't have one of the ingredients attributed to him.

coherently about the question of why there is something rather than nothing.[11]

Regarding the self, Buddhist tradition emphasizes meditation over ritual. The aims of meditation are self-awareness, self-knowledge, self-control, and self-transformation. Many Buddhists are master phenomenologists. In the West, Augustine's *Confessions* and Marcel Proust's *Remembrance of Things Past* stand out as great works that take a minded life as their object and show their authors' patient, nuanced introspective skills. The philosophical tradition that the West officially dubs "phenomenology" began in the nineteenth century with the work of Edmund Husserl, a student of Franz Brentano and a teacher of Martin Heidegger. In his *Ideen* (*Ideas*), Husserl proposes a discipline, a set of techniques for providing reliable phenomenological reports. At the same time, the founders of "scientific psychology"—Wilhelm Wundt, most prominently—were training introspectors to give reliable reports about their experiences in psychophysical experiments. Phenomenology or introspection is not, according to its Western practitioners, something one can just do, at least not if one wants to see things clearly and correctly. These things take discipline.

For many centuries before Husserl and Wundt, Buddhist meditation masters similarly stressed that self-knowledge or what is often called "mindfulness" required training, including something Husserl stressed, working to eliminate preconceptions about how experience is supposed to be as a condition of discovering how it really is.

What do Buddhists experience when they engage in mindfulness? They experience what William James, himself a great phenomenologist though not an official member of the European school of phenomenology, called the stream of conscious experience.

Nowhere does any Buddhist phenomenologist see, experience, or discover a permanent self, an abiding mind's "I." Experiences flow, and

---

[11]Confucianism is another tradition that is not very theological by our standards. In Confucius's *Analects*, as well as in other Confucian texts, references to "heaven" and sometimes to "God" appear. But most Confucian scholars recommend reading these words as referring to "everything that exists" or alternatively to "the we-know-not-what, but that which is greater."

they flow sensibly or coherently, bound by memory and a feeling of accumulation. The posit of a self, or better, a person makes sense if we think of mindfulness, introspection, or phenomenology as an activity performed by a changing system—a continuous and for now relatively self-contained changing system—on itself. Over time, both the "observer," the system, and what is "observed," the system's experiences, are ever-changing. Nothing stands still, nothing stands in a stable relation of observer to observed. We are just as Heidegger insists, beings-in-time. Our being is essentially temporal, wrapped in and modified over time.

The Buddhist can make sense of the projects of self-scrutiny, self-knowledge, self-control, and self-transformation without supposing that there exists a self over and above the stream of consciousness. Experience never reveals a "see-er" independent of some act of seeing, even when what is seen is myself thinking. These things are inextricably codependent. The illusion that there is an independent and permanent "I" that sees things and orchestrates experience, and to which all experience belongs, is undermined by the phenomenology of experience. The alleged independent "I" that might seem to be the permanent "see-er" can itself at the very next moment become an object of introspection. What appeared as subject can in a moment be made into an object. So the "I" that might seem to own an experience can itself be examined. "I see that I saw the orange a moment ago." Now there are two "I"'s. Which one is the permanent self that exists independently of the stream? The Buddhist answer is neither. There is no permanent self revealed phenomenologically. "I" is just a word that we use to indicate that the ever-changing subject of experience is having some experience or other. The magic indexical "I," the related pronouns "me" and "myself," as well as our possession of an unchanging proper first name, in my case "Owen," contribute to the illusion of a permanent, stable, and immutable self—an illusion that mindfulness undermines.

I can only guess why Buddhism had the resources to see through this illusion while Western philosophy and religion have held onto it so resolutely. In his writings, and in my conversations with him, the Dalai Lama expresses a genuine open-mindedness and antidoctrinaire sensibility. He is frankly and openly worried about defenders of any religious orthodoxy, seeing in their leaders and followers a dangerous hubris that engenders intolerance and displays a close-minded confidence that we already know the answers to all of life's mysteries. Buddhism is open generally, but interestingly open to science. The Dalai Lama is very interested in what evolutionary biology, cognitive neuroscience, and particle physics have to teach us about the nature of mind and its relation to the world. I don't know how he would respond to work in these areas if they were not so friendly to the views he already holds.

What I can say, however, is that Buddhism has traditionally avoided overreaching, in a way that the manifest image that we abide in the West has not. Herein lies a general lesson for constructing a manifest image. Buddhism has humbly constructed its belief system around what can be observed in the ordinary world—where cause and effect relations are ubiquitous, everything is impermanent, suffering can be alleviated to some extent by compassion, and humans possess a continuous streamlike mental life for as long as they are alive. What Buddhists call "bare attention," which is hardly bare, involves scrupulous attention to the way things—both the world and our own minds—really seem. Bare attention allows things to reveal themselves as they seem when carefully attended to, and it yields and grounds these observations. But it provides no grounds for speaking about much more than appearances. More extensive knowledge, deeper knowledge, awaits the unfolding of history and the discovery of methods that allow us to reliably infer more than what bare attention reveals. History has produced such methods, and they include science. Having said little beyond what ordinary experience reveals has positioned Buddhism, possibly uniquely, to meet science with comfort.

In the West, aided by a perennial philosophy that for much of its history was theology's handmaiden, we have conspicuously overreached in constructing the main components of our manifest image. There is very little about which the Western manifest image is not opinionated. The resistance to Copernican views of planetary motion, to Darwin's theory of evolution, and in our own time to the view that mental life is the life of an impermanent, embodied organism with a brain arise because contrary positions were already held as unquestioned doctrines before these theories were proposed. This is perhaps why great thinkers like René Descartes and Thomas Reid, when they engaged in introspection, saw something that no good unbiased phenomenologist could see—a permanent, simple, and indivisible self. Not only is rigorous phenomenology underdeveloped in the West, but when it has been practiced it has been distorted by prior, overreaching, and doctrinaire views.

In this chapter I have devoted my energies primarily to the destructive task of showing that the part of our manifest image committed to the belief in nonphysical substances—God and mind—is misguided. In particular, the belief that persons are constituted essentially by their possession of a nonphysical, permanent, and immutable soul is untenable. I have shown as much patience and care as I have been able to muster in saying why this belief is wrongheaded, because it is such a deep part of the way many people, including many I love and respect, think about themselves. The belief can't be displaced simply by saying that it is old-fashioned, and certainly not by ridiculing it. I have done my best to show that the arguments for the self as conceived by the manifest image, arguments that are seldom discussed and analyzed but are taken for granted as part of the background, are bad arguments. There is no self that is constituted by an immortal soul, and God is something about which, if you have good sense, you will resist speaking, especially with an air of confidence. Nothing sensible can be said.

But I do have a positive view of the self. I now proceed to say what selves really are, given that they are not what we take them to be.

SIX

Natural Selves

> The nucleus of the "*me*" is always the bodily existence felt to be present at the time.
>
> —*William James, 1890*

> [T]hough each pulse of consciousness dies away and is replaced by another, yet that other, among the things it knows, knows its predecessor, and finding it "warm" . . . greets it saying "Thou art *mine*, and part of the same self with me. . . ." It is this trick which the nascent thought has of adopting the expiring thought and "adopting" it, which leads to the appropriation of most of the remoter constituents of the self. Who owns the last self owns the self before the last, for what possesses the possessor possesses the possessed. . . . [T]he thoughts themselves are the thinkers.
>
> —*William James, 1890*

I BEGAN THE LAST CHAPTER with this claim and these questions: Nothing is more important to me, my friends, and loved ones, than who I am. But what is this me, this self that is so very important? Suppose there is a self, suppose that the world is filled with selves, what are they, where are they?

I spent most of last chapter saying what the self cannot be. It cannot be the soul, because there is no such thing. If there is something properly called "the self" it is not simple, not indivisible, not nonphysical, not immutable, not immortal. But now the reader is owed a positive conception. I need to say more than that the positive view I recommend can be found in Buddhist texts. I could add that the positive view can also be found in Derek Parfit's magisterial *Reasons and Persons,* where Parfit notes the strong similarity with Buddhist views, in William James's *Principles of Psychology,* as well as in Varela, Thompson, and Rosch's *The Embodied Mind.* But in doing this I will not have said what this view is. I need to lay out the positive conception.

Since I already feel a bit like the Grinch who stole Christmas, I hesitate to say what I am about to say. But here goes. One might have the thought that the critique of the idea that the self is the soul already draws into view the shape of a partly consoling positive proposal. We might take all the things that I claim the self is not and think the positive view goes, in broad strokes, as follows: The self is complex, it is divisible, it is physical, it changes over time, it is mortal. But this conclusion cannot be taken for granted. It does not follow automatically that there is a self constituted by the positive features one gets by following out the implications of denying certain other traits. From the facts that Santa Claus is not fat, does not have a white beard, and does not live at the North Pole, it does not follow that Santa Claus is thin, has a different colored beard or is clean shaven, and lives someplace other than the North Pole. One must allow that Santa Claus has no attributes at all, because he doesn't exist.

It is possible, too, that there is no such thing as the self. Perhaps the concept not only cannot be made sense of in terms of an incorporeal and immortal soul, but cannot be made sense of at all.

I don't think this, nor do I think that commitment to the scientific image implies that belief in the self is an illusion. But I point out the possibility because there are those who do think this. David Hume is sometimes read as thinking that there is no such thing as the self, although I believe Hume is better read as thinking that his self is something, but not something constant.

There are two sorts of folks who think that nothing real answers to the description of the term "the self." Some are postmodernist thinkers, the others are scientific reductionists. The first sort think a self is something created by wordplay or fictional construction, similar to King Arthur or the tooth fairy. There are words for these things, but they refer to nothing real. The second think that science admits into the ontological table of elements only what is really real, and what is really real is revealed only by the basic sciences. Physics, the most basic science, makes no mention of selves. So selves don't exist.

The industrial-strength versions of the postmodernist view and what I call the reductive or scientistic view—or what one philosopher of mind, Rob Wilson, calls "smallism"—go like this:

Postmodernism: Descartes, the father of modern philosophy, was a modernist. He believed, among other things, that the self exists, indeed that its existence is the most certain thing.[1] Furthermore, the self is the soul. Postmodernists, skeptical of claims to certainty, believe that persons engage in language games out of which emerge fictional objects, the self among them. The self is a product of narrative construction. Individuals take memories (these are inherently constructive) from events in their lives; we start with materials reconstructed from what we have done, things that have happened to us, add a healthy dose of self-serving interpretation, color the story with all manner of projection of how we would have liked to be and how we would like to seem to ourselves and to others, spin the ingredients in accordance with an interpretive model—a psychoanalytic model, a tragic or comic model, a Buddhist or Christian model—and *voilà*, a self emerges.

But the self that emerges is just a fictional entity, a narrative construction. One might insist that surely the narrative self is different in kind from fictional objects like King Lear or a unicorn. To which the postmodernist will reply "no." There are real kings, fathers fixated on the love of their daughters, and horses and animals with horns. Fictional constructions or characters start with these real materials,[2] but the reason they are fictional is that they contain too many additives.

---

[1]Here I am interpreting Descartes as equating what he calls "mind" with the soul. If we worry, as I did in the previous chapter, that Descartes's picture of mind doesn't explicitly admit a proper soul, an immutable container of experiences, that makes experience possible, we will question his right to this identification. We will either need to expand his conception of mind to include "the right stuff," assuming that it is already presupposed, but insufficiently articulated, or we can add a third ingredient to Descartes's dualism of mind and body and make it the view that we have three parts: a soul that houses the mind and a body.

[2]The usual route to skepticism about real ingredients, actual events, and the like is to say that whatever we think or speak of is theory-laden—perhaps solely by virtue of the fact that even our ordinary everyday language embodies a "theory," for example, it assumes that there are physical objects. Of course, this sort of global fictionalism makes the claim

Just as a food is not "organic" if chemical herbicides were used to grow it, a concept is fictional if fantasy and imagination play more than a minimal role in its construction. Self-construction and self-description start with real characteristics of real people—oneself and others. But what emerges is invariably too colored by interpretation, hopes, expectations, and a style of narrative composition (a "theory" of the self as guided by unconscious Freudian impulses, a self etched with the scar of original sin, a self who lives by the hand of Fate) to differ in kind from these other fictional constructs.

There is something right about this view, but it is overstated. It overemphasizes the degree to which self-narration involves fictional aspects, which it undoubtedly does, while underestimating the constraints that keep self-narration on track. In normal cases, our concept of our self remains sufficiently grounded in reality that it refers to something in the vicinity of who we really are.

Smallism: The person gripped by the reductive or scientistic impulse comes to reject the existence of selves by a very different route. She seizes on the alleged insight that reality is best captured by the basic sciences, physics in particular, to conclude that all there is is what physics says there is. Everyone agrees that everything obeys the laws of physics. Electrons, protons, quarks, muons, and gluons appear in the laws of physics, but these things do not appear anywhere in the "laws" of psychology, sociology, anthropology, or economics. Meanwhile, "mind," "brain," "society," and "culture" are nowhere mentioned in the laws of physics. Nor, of course, are "selves." Since physics reveals what is, what is real, it follows that what it does not mention does not exist. So much for most of what you believed in.

Some of my best friends were once eliminativists (the most familiar breed of "smallists," who believed that folk psychology is bunk awaiting replacement by mature neuroscience), but they have almost

---

that the self is a fiction somewhat less interesting, for my geraniums in my garden are also fictions in this view. Global fictionalism is just the view that everything is a fiction. One plausible idea might be to accept the points about the ubiquity of interpretation and theory-ladenness, but then demand that some help be offered to figure out matters of degree, so that we can distinguish among things that are real fictions, but not seen as such, and things that are described by importing interpretive and theoretical elements in unobjectionable ways.

all seen that there is something seriously mad about the view. Most mad-dog eliminativists have quietly transformed themselves into advocates of the sane view that talk at the level of mentality (psychology) ought to be constrained by realistic assumptions at the level of brain science. This will prevent posits of such things as the soul, and it will rein in loose or magical theorizing of the "and then a miracle happens" variety.

The postmodernist makes the mistake of underestimating the degree to which self-description is constrained by forces that keep self-narration on track. The "smallist" makes the mistake of thinking that, because everything is constrained by the laws of physics, therefore physics alone describes the really real. But most everyone now believes that there are emergent natural properties that, despite being obedient to the laws of physics, are not reducible to physics. There are levels in nature. If the "smallist" had her way, not only mind, society, and self would go but also bacteria, neurons, brains, plants, and animals.

Trust me, the world is safe for the self. Neither postmodernism nor "smallism" provides compelling grounds for thinking otherwise. I assure you that there will be no *National Enquirer* headline that announces "Literary Critics, Cultural Studies Scholars, and Scientists Prove That There Is No Self." Or if it does, it will be as believable as most of the *Enquirer*'s other grave pronouncements.

Self-Emergence in the Material World

When I was conceived I did not possess a self. My father's sperm was selfless, as was my mother's ovum, as was the conceptus that became me. In a sense, even then, I was something rather than nothing, although I was not remotely possessed of selfhood, or me-ness, but that conceptus had the potential to achieve it. I have actualized this potential.

The story of how a conceptus becomes a fetus that becomes a viable human infant is well understood. But how a human infant acquires a sense of self, becomes a person, establishes an identity, and eventually gets in touch with this identity is less well understood. The sequencing of the human genome may add to our understanding of these processes, especially in the early stages of the development of the self. But genetics is not likely to be a major player in explaining the self.[3] Selves emerge in embodied organisms living in multifarious social environments. And despite much hoopla, neither sociobiology nor the new evolutionary psychology displays remotely adequate resources to explain, on its own, the varieties of social worlds that exist and that contribute significantly to self-emergence. Nonetheless, a plausible theory of the self is coming into view, thanks to the combined efforts of philosophy, child psychology, cognitive neuroscience, evolutionary psychology, sociology, and anthropology, as well as work in literary theory on narrative.

Daniel Stern, a child psychiatrist, sees the sense of self as a primary organizational principle in all areas of development. In broad strokes, his theory has the following main components. First, in the first eight weeks of life, "the infant can experience the process of emerging organization," which Stern calls "the *emergent sense of self*." Second, during the period of "two to six months, infants consolidate the sense of a core self as a separate, cohesive, bounded, physical unit, with a sense of their own agency, affectivity, and continuity in time." This core *self* involves a clear sense that one goes on being. Third, between seven and fifteen months the child develops a firm sense of himself as a subjective being, and he becomes increasingly adept both at conveying his inner life to others and at interpreting their inner states. This third stage is the first that can plausibly be said to involve metacognition, a reflexive pose—awareness that I experience myself in "this" or "that" sort of way. The project of solving the problem of other minds is motivated in part by the fact that the child is not only achieving greater autonomy but is also using her newfound psycho-

[3]One area where the human genome project will likely help greatly in understanding the self is in psychopathology. Certain types of mental illness, schizophrenia, for example, occur absent any traumatic brain injury. Some schizophrenics don't feel their own consciousness as theirs, they experience others thinking for them or inserting thoughts in them. Insofar as schizophrenia (and other forms of mental illness) has genetic roots—but I hasten to point out there are interesting viral theories out there—the human genome project will help us to understand its causes, possibly stop its onset, or enable us to cure it. See G. Lynn Stephens and George Graham's wonderful book, *When Self-Consciousness Breaks: Alien Voices and Inserted Thoughts,* for a rich exploration of what it is like to be a person who does not experience her experiences as her own.

logical savvy to solidify and deepen various interpersonal unions. We are codependent rational animals.

Finally, during the second year, we see the emergence of language and the ability to self-represent. The degree to which the child uses language for self-representing is highly variable across cultures and individuals. The emergence of the verbal self is not the marker that the self is beginning to emerge but a marker that the process is well under way.

The cognitive neuroscientist Antonio Damasio distinguishes among the proto self, the core self, and the autobiographical self. Each depends on its predecessor: The autobiographical self rides on top of the core self, which rides on top of the proto self. The proto self is unconscious and constituted by "interconnected and temporarily coherent collections of neural patterns which represent the state of the organism." We can imagine a lobster or a fetus in the first or second trimester as possessed of a proto self. The core self is nonverbal or preverbal. Any organism possessed of a core self is a subject of experience. The core self is conscious. Dogs, cats, and human infants have core selves. The autobiographical self requires maturation of autobiographical memory, which probably requires, at least in the human case, that one have language or be in the process of acquiring it.

I do not want to insist that autobiographical memory requires full-blown linguistic competence, or that self-knowledge is necessarily couched in a natural language.[4] Autobiographical memory does seem to require something like conceptual structure, the capacities to form and hold concepts consciously, and in the human case these capacities appear to be inseparable from linguistic capacity. But I don't mean to suggest that only creatures capable of acquiring humanlike language can possess autobiographical selves. Chimps, bonobos, and gorillas all have trouble learning human language—despite noble

[4]This is a matter of considerable debate. Many philosophers think that all thought requires concepts and that all concepts are linguistic, and thus that creatures, including human infants, lack thought. A stronger view still is that experience requires concepts, which requires language, and thus that nonlinguistic creatures do not have anything that can properly be called experiences. The big problem, especially for the stronger version of the view, comes from the overwhelming evidence that animals experience pain.

efforts to teach them—but I have no doubt that they have concepts that they utilize in self-comprehension. Furthermore, although we couch our self-descriptions in language, our selves are not exhaustively captured in what we can say about them. There are many things I feel, many ways I feel things, including things about my self, that I can't put into words.

In any case, Damasio's theory fits comfortably with Stern's but has the added advantage of describing the neural circuitry for each developmental stage. The proto self requires a body that sends signals (via chemicals in the bloodstream) to the basal forebrain, hypothalamus, and brain stem, which cause the release of certain neurotransmitters in the cerebral cortex, thalamus, and basal ganglia. With these pathways the organism has the right stuff to unconsciously represent various states of its own body, including its relation to certain external objects. The core self requires the apparatus of the proto self to be up and running, and, in addition, for the cingulate cortex, the thalamus, some parts of the prefrontal cortex, and possibly the superior colliculi (twin structures in the back part of the midbrain) to get into the act. Once these brain areas come into play we have a genuine subject of experience. The organism feels things and feels itself feeling things.

The autobiographical self, what Damasio also calls "extended consciousness," requires that we be able to hold memories over time. A creature with autobiographical memory partakes of more than just passing states of consciousness. It can see and hold its life, or segments of its life, in view. All the evidence points to memory storage operating in a distributed manner. My knowledge of what a chicken looks like is distributed over neurons in one cortical area, what it tastes like in another, what it sounds like in yet another. Writing the last sentence just activated my memory of several "Why did the chicken cross the road?" jokes. These are recorded in a still different site.[5] But the fact that these jokes came to mind suggests an important

[5]The standard view is that mental representations, and thus memories, are stored in a distributed fashion. So we should not think of each representation having its own and only its own site. That is, a neuron that is part of the "net" that holds my memory of what chicken tastes like might well be part of a network, no doubt composed of distinct neurons, that makes up my chicken joke memory.

feature of the system. Get me thinking about one aspect of a chicken, and all the relevant related records will be activated to some degree.

In any case, the point is that all long-term memory is distributed and multisited. Autobiographical memories are stored in various early sensory cortices and activated by convergence zones in the temporal and frontal higher cortices, as well as in subcortical areas, such as the amygdala, which are particularly important in experiencing and remembering how certain experiences feel and felt. Consciously entertaining specific autobiographical memories requires activation of "working memory" in the prefrontal cortex.

Even if you know nothing about brain anatomy, and thus didn't understand a word I just said, you will understand what I was trying to explain: All that is required for self-emergence to take place is for an organism with a body that houses a certain kind of brain to live in the natural world with other similarly embodied creatures. Madonna, the singer and metaphysician, had it right: "You know that we are living in a material world and that I am a material girl."[6]

The Magical Indexical "I"

Infants have core selves. There is something it is like to be a newborn—something it feels like to be a newborn for the newborn. It is credible to describe "what it is like to be" an infant as involving an inchoate sense that something is happening here, where "here" gestures elusively to what the newborn will someday come to designate in more transparently self-referential terms. According to the psychologist Jerome Kagan, utterances deploying self-referential terms—

---

[6]Sometimes it helps people who can't imagine the brain being enough to house the self to run some numbers by them. Neurons are the basic cells of the brain. There are at least 100 billion neurons in your brain, at least 10^{11}, possibly as many as 10^{14}. These neurons communicate with other neurons at junctures called "synapses" (by way of lines of input and output known as dendrites and axons). If we assume as is plausible that each neuron has synaptic connection with 3,000 other neurons, then this is on the order, again conservatively, of 100 trillion synaptic connections. The axons in your brain alone if connected and straightened out might well reach from New York to San Francisco. I haven't even mentioned glial cells, which outnumber neurons by about three to one, and appear to be importantly implicated in creating synaptic connections. You get the drift. Indeed, you are the drift.

"I," "my," "mine," "me," "myself," or the child's name with a verb ("Sam eats")—are absent until about 19 months but are common by 27 months. Around their second birthday, children are competent at locating themselves in the causal nexus, in relation to external objects, as sources of action and as fountains of desires. Even when self-referential capacities arrive we need infer nothing very exotic in terms of self-awareness. Most self-referential utterances serve simply to locate the agent and express her thoughts and desires as hers. Yet this is no small accomplishment. By age two, the child has acquired some sort of self-concept. She can clearly mark off herself and her individual actions, beliefs, and desires from the rest of things. But when she uses her name or words like "I," "me," and "myself," she is not, in any deep sense, saying who she is or what she is like. In part, this is because these words serve primarily indexical functions. They serve to point rather than say who one is. Furthermore, when the child first acquires self-referential capacities, there is no deep way that she is, at least not yet. That requires a narrative, and at two she has neither stories to tell nor the ability to tell them.

When I was 16, I had my first romance. I fell in love with Barbara. I told her so. I also said to some of my best pals, "I am in love with Barbara." By then I was something of an aficionado at self-description. The teenage years are a peak period for private and public self-scrutiny. At the same time that one is increasingly unrevealing to one's parents, one is increasingly self-revealing to oneself and one's friends. True to the nature of other teenage selves, my self was ridden with angst, aided and abetted by listening nonstop to early Dylan recordings. At 16 I possessed a rich autobiographical self and a sea of very confusing emotions.

Or so it seemed at the time. In retrospect, my "then-self" seems thin and underdeveloped compared to the self I am now. Perhaps this is so, or perhaps my memory for my "then-self" is dim. In any case, by the time one can sensibly make such an utterance as "I am in love with Barbara," the "I" expresses more than the simple location of where the action is. It is also taken, by both my friends and me, to refer to Owen at 16, this person with a history, with a rich self-concept. By the time some individual is 16, "I" still refers to where the action is, it hasn't lost its indexical function, but it also refers to this self that is now clothed in sixteen years of experience, who has accumulated knowledge, traits of character, and emotions, including, we hope, some sense of what love is.

But language can trick us. We may think that "I" refers to more than a being-in-time with depth and rich texture, to more than a genuine character with certain stable personality traits and aspirations. In one sense it does. When I said "I am in love with Barbara," I was expressing only one very salient fact about myself. I could have said much more. My friends also knew a lot about me. Whenever a person reports that he is in love he assumes that his listeners know what "love" is, and, explicitly or implicitly, know enough about him to make sense of the report. My listeners' (and my own) knowledge of my personality and my background might make us prone to think that "I" picks out something even deeper than the self I am. We might think that the "I" refers to more than Owen at 16, and in fact refers to the "I" that accompanies Owen through life, Owen's transcendental ego.

We need to beware of falling into this trap. It is hard, as we saw in the previous chapter, but it can be done. Think of Owen at 18, when he met and fell in love with Amy. He then felt and said, "I am no longer in love Barbara." "I" once again pointed to where the action was and also referred to a person, me. It referred to Owen at 18, the person psychologically continuous with Owen at 16, but possessed of two years of additional experience. Nothing deeper need be inferred than that I changed some due to my additional time on earth. So much about me remained the same during those two years that I (or my loved ones and friends) might have fallen for the idea that some immutable ego accompanied me on the journey. It didn't. There is no ahistorical "I," no transcendental ego, no immutable soul that accompanies me on this journey. I am an embodied, historical being all the way through.

Indexical Therapy

Many contemporary philosophers are still gripped by what they think is the deep mystery of why experiences are had uniquely by the subject of them. Several very intelligent philosophers, including some good friends, have received tenure for publishing long books on how subjectivity is possible, how brain tissue can cause consciousness. The problem is surprisingly easy to solve—at least from my armchair. Experiences are unique in having what John Searle calls "first person ontology." They appear in a unique way to the subject who has them. What I called "subjective realism" in the first chapter is true of conscious animals. Why is that? Nothing mysterious. Each individual has her own and only her own experiences because only she is connected directly to her own nervous system. End of story.

That said, the magical indexical "I" is something of a troublemaker. Owen at six felt humiliated, deeply embarrassed, as he heard his classmates laughing and realized that they were laughing at him because he was holding his Thanksgiving picture upside down. Suppose that at that time he said to himself, "Oh God, I am so embarrassed." What was this "I" that is having the thought? "I" is how we denote the biological and psychological continuity of our unique first-person stream of consciousness. When we say, at 52, such things as "When I was 8, I stole the contents of the Easter collection box, in which I was saving money during Lent for the poor, in order to buy myself a new hamster," "I" refers to what this subject of experience now did then when I was someone different but continuous with the person I am now. I was different then in many ways—limited experience, lacking training or wisdom in prudence, and excessively motivated by short-term satisfaction. When the smashed poor box was discovered under my bed, my parents were not pleased. Explaining that my pal Reggie Sutherland and I *both* thought that our need to buy a hamster (98 cents in those days) trumped the needs of the poor did not help our side.

When we catch ourselves doing things or when we remember having done things in the past, we rightly apply the magical indexical "I." But the "I" is not something permanent that accompanies all experiences. There is no credible sense in which the "I" who robbed the poor box to buy the much-needed hamster is the same "I" who is writing these words. Yes, I am psychologically and bodily continuous with that poor-box thief of 44 years ago. The temptation to posit a permanent "I" is strong and if we yield to it, it can foster the illusion of the "mind's I," a permanent soul or ego that stands behind all experience. This is a bad idea. James's idea is better: "the thoughts themselves are the thinker." But even this is not the best way to put things. We think with our brain. We think with thoughts. But it would be best to say, "the person himself is the thinker." A person is what thinks—a whole embodied being with a history.

This idea makes sense if we recognize that the conscious stream is embodied as a stream of self-awareness. Furthermore, we, persons, that is, possess fancy reflexive capacities to think thoughts about where we, by virtue of embodying and possessing our stream (it is embodied of course), have been, what is now being experienced, and where we are headed. When we find it important to think or report what we are experiencing or doing we make use of the magical indexical "I." This is the best way to locate in time what is happening, what we are feeling, thinking, and doing. The mistake is to think that the self-locating thought "I" now refers to the same "I" it refers to when I think about my past—to the exact same self that performed some act in the past, like stealing the money to buy a hamster.

We can overcome the illusion of the "mind's I" by thinking about other pointing terms. Like indexicals such as "I," "me," and "myself," demonstratives such as "this," "that," "here," "then," "now" also serve to point. Often these words are used to point to different things—that azalea, that house, that rabbit. Other times they are used to point to the same thing—*that* azalea on May 3rd and *that* (same) azalea on June 25th. The reidentified azalea is the same in the sense that it is physically continuous with the azalea pointed out earlier, but it is not exactly the same. Perhaps it has flowered and grown. When we say "now" we always, in a sense, point to the same thing, the present moment. But no one thinks "now" refers to anything more than the present moment, which is always different. No one thinks, for example, that "now" refers to an abiding moment that reappears again and again. One reason is that the word "now" is commonly used in sentences that refer to external states of affairs: "It is raining now"; "Now is the time for all good men to come to the aid of their country"; "Now it is time for bed."

These sentences draw our attention to these external states of affairs, which we experience as different and distinct from each other and from us. The indexical "I" performs its magic seduction of fostering the temptation to posit a soul, what William James called an "arch-ego," partly because it always indexes what is first-personal and inner. As James said, each experience in the stream comes with a feeling of warmth and intimacy—a feeling of being mine. "I" always points (at least in part) to this subject of experience, this subject who experiences in a most intimate way the journey that is my life.

Despite this difference, "I" is best understood as functioning, first and foremost, just like all other pointing terms, as a term that indexes me at some point in time. However, by the time I am someone, have a life and thus a self, as it were, "I" not only points to me now but often implicitly gestures at who I am. When I say, "I am having some trouble with Ben's move to attend college in San Francisco," I am pointing to me, to this subject of experience. But "I" also implicitly refers to this self, the father of Ben, who loves him deeply, who misses him greatly, who is concerned about geographical distance engendering personal distance, who wonders and worries about how Joyce and Kate will respond to his move, and who is concerned that Ben will devote more time to doing music than to school.

The fact that "I" uttered today seems just like uttering the word yesterday is, first, because the conscious stream is sensibly and subjectively continuous and, second, because in the normal course of things we change very little from day to day.

"Here," "there," "this," "that," "then," "now" are thin demonstrative markers. They only point or mark. "I" is often used as a thin indexical, such as when I tell an anonymous grocer that "I would like a pound of grated Parmesan." But often, especially in most important interpersonal relations, "I" functions as a thick indexical. It points or marks, but it also expresses and evokes the sense that what is happening is happening not only to me but to this individual with a particular identity, with a unique personality. In order to really understand what my "I" utterance means, I want and expect you to understand me for who I am.

The Hippie Without a Cause

My playful 14-year-old nephew, Pete, teases that he is going to write a biography of his Uncle Owen, to be entitled "Hippie Without a Cause." Let me tell you a bit about myself, not in Pete's terms but in my own.

I was born on January 30, 1949, in Bronxville, New York, to Virginia Lyons and Owen Flanagan Sr. They were 21 and 22, respectively. I weighed 9 pounds 4 ounces and was 23 inches long (I now weigh 185 pounds and am 71 inches tall). Dad finished college at night after the war and we moved to Puerto Rico when I was one year old so that Dad could take a job at a New York-based accounting firm with several major clients in San Juan. Except for my current height and weight, these facts are, as Spinoza would say, "knowledge by hearsay."

I remember a few things about our four years in Puerto Rico. I remember the mosquito netting that Mom covered me with each night and the smell of the insecticide she then sprayed to keep the bugs away. I remember Marianna, our maid, who checked for scorpions in my shoes each morning. I remember Marianna telling me "Owen sito, que guapo" ("little Owen, very handsome"). I remember her taking me to a park and buying me Hershey's kisses.

We returned to the States when I was five, with my sister Virginia, 18 months younger than me, in tow. We bought a house in Hartsdale, New York, and I entered Sacred Heart Grammar School. Because Sacred Heart had no kindergarten, I had to go into first grade, where I was younger, smaller, and more timid than the other boys. I was very frightened by public speaking, to the point that I would not ask for permission to go to the lavatory to pee when I desperately needed to, squirming instead until the bell rang. In second grade, I was asked to stand in front of the class with several other kids to show our Thanksgiving pictures to the rest of the class. Everyone laughed because I held mine upside down. I was mortified. During grammar school we spent a great deal of time practicing air raid drills, taking refuge under our desks in case nuclear weapons struck New York City.

FAST FORWARD

After eighth grade I went on to Archbishop Stepinac High School in White Plains, New York. By then I had four younger siblings, and a fifth would join us in my senior year. I was an excellent student. I loved Greek, Latin, English, and Mathematics especially. And I have already told you I fell in love with Barbara. I had my first two-party sexual experiences with Barbara.

During high school, the Cuban missile crisis of 1962 helped solidify a thought I had often had under the desk during air raid drills in grammar school. Even if communism was bad, we would be communists if we had been born in a communist country (no one, it seemed, acknowledged this obvious fact). I came to disrespect most adults for not examining their beliefs and especially for not seeing that many of their deepest commitments were due to utter accidents involving the temporal and geographical location of their births. I thought of most adults as unreflective and lemminglike. I trusted my teachers to teach me what they knew, but I came not to trust political or religious views that were thoughtlessly handed over. My rebellion was quiet and polite, largely carried out in the privacy of my heart with considerable fear, anxiety, and a vivid and abiding sense of loneliness.

Dad noticed that I had lapsed, or was in the process of lapsing, from the faith. He gave me an abridged copy of Aquinas's Summa Theologica, which I read with much enjoyment. I saw that Aquinas's five proofs for the existence of God were failures but I was, as I suspect Dad knew I would be, taken by the seriousness of his efforts and by what I later understood was philosophy. The trouble was that one of my main thoughts after reading Aquinas was to wonder if you could actually get paid for doing this kind of work. The idea that young Owen would attend the University of Virginia or Michigan and become a lawyer like my Grandfather Lyons and my Uncle Austin, or become a CPA and join the firm of Smith and Flanagan, started to recede in my mind as thoughts of becoming a philosopher gained ascendancy.

When I arrived at college, my first day of philosophy class opened with these (to me) unforgettable words: "Plato posits

The Good." I did not know what "positing" was and I had never heard the definite article stand before the word "Good"—which I rightly heard as capitalized. But I was thrilled, captivated, and hooked. I eventually decided to take a Ph.D. in philosophy rather than psychology. It was something of a close call. But in those days experimental psychology used rats, and Owen didn't and doesn't do rats. I was also gripped by the idea that the philosopher's job was to question unquestioned beliefs, something I was already by then pretty practiced at.

My philosophical passion was rooted in an attempt to address the religious questions that still, in a disguised or transformed way, absorbed me after my lapsing. Why is there something rather than nothing? Why am I one of the things that exists? If there is no God, then is everything allowed? What is goodness? Why be good? What is the meaning of life? What is mind? What is a person? What is the self? These are the same questions that absorb me today.

In college I was a hippie. I grew a beard and wore my hair long (as I still do). I smoked marijuana religiously, opposed the war in Vietnam, and organized my personal and political values around the mantra of "peace, love, and happiness." My daughter Kate calls me "an aging hippie." I guess I am.

FAST FORWARD

After I received my Ph.D., I took a job at Wellesley College in 1977, where I taught for 16 years before moving to Duke University in 1993. I love teaching, although I continue to find it nervous-making. I also love to write, although I also find it nervous-making. Truth be told, I find life itself nervous-making.

In 1980, I married Joyce Knowlton Walworth. Ben, our son, was born in 1981 and Kate, our daughter, arrived in 1983. I lost my wonderful mother, at age 60, to cancer in 1987, and my brother Peter, only 21, in a solo drunk driving accident six weeks later. A depressive cloud enveloped me about two months after Peter's death and I sought treatment by a psychiatrist for my anxiety and depression. In 1993, just before we moved to North Carolina, I was diagnosed with a pituitary tumor after noticing that I had lost my libido (completely) and that my ectomorphic runner's body was putting on weight. I went on medication, which I still take, to control the pituitary problem, keep the tumor small, and regulate my hormones. The medication tells my brain to regulate dopamine production, raise testosterone levels, and reduce prolactin production (guys don't need much of this). It works. And although I might some day need surgery to remove the growth (especially if it starts hitting my optic nerve, with which it is now quite friendly), I don't yet need to have any brain parts removed. Knowing a lot about the brain means knowing that it is best to keep the surgeons at bay for as long as possible.

But the medication did not work right away, and finding the right prescription regimen for my anxiety, depression, and my new friend, my tumor, proved formidable. In 1994, one combination of meds sent me into a manic episode. I behaved irresponsibly and acted out-of-character for a time. I lost a certain amount of self-esteem and self-respect, played games with myself and my loved ones, and faltered at the things and relationships I cared most about.

My father died suddenly of a stroke at age 70 in 1996, and my sister Virginia died of a sudden heart attack in 1998. At Dad's funeral my Uncle Bill (Mom's older brother and Dad's best friend) toasted my father with a hefty glass of gin, mentioning that he was about to divest his gin stock, due to the impending decline in worldwide consumption that would be caused by Dad's passing. We all had a hearty laugh, despite the fact that active and inactive alcoholism abounds in our family. I gave the eulogy at Virginia's funeral. She was a loving and dedicated mother but extremely mischievous as a child and teenager. I told stories about both sides of "Vee." Joyce and I separated in 1998 and divorced. According to a not implausible version of the story of our breakup, my tumor, my medication, my personality, and my behavior—on the long road to my return to my pretumor self—took too much of a toll on our relationship. I have many regrets about how things turned out, including how I was that contributed to the failure of our marriage.

Nowadays, I am something of a loner, sometimes lonely, despite being considered by many to be lively and outgoing, even charismatic. I see why they think this. But deep inside there remains a very shy, somewhat withdrawn person. I adore Ben and Kate and the feeling is mutual. And I still love Joyce deeply. I have been a novice Buddhist practitioner for awhile now. I meditate each day, work to moderate and modify destructive feelings and emotions, try to live simply and without acquisitiveness, and do my best to practice an ethic of universal love and compassion. In general, I try, as best I can, to accept what I can't change, to enjoy the here and now, and "to strive on with awareness" as the Buddha recommended. In addition to my family, my work, and my flower gardens, world travel is a major passion. I see myself as a sort of amateur anthropologist. Travel and exploration, making new acquaintances far and wide, help me to be a better philosopher and to feel happy, more complete. Perhaps I am an aging hippie, as Kate would say, but I am fortunate to be passionately gripped by causes, projects, and plans. There are people whom I love deeply, relationships to nourish and grow, and many, many things left to do. I am lucky.

I PROVIDE THIS LITTLE SKETCH, this Monarch Notes version of my life, partly to say some true things about myself so that you can know a bit about me. But I also hope this story might serve as a guide to some things I believe to be true of selves generally. I didn't consciously or actively intend to tell the story in a way that would do this. I simply expected that, as a story of a person's life, it would reveal some normal features of people's lives. And indeed it does. First, like you I am a being-in-time. My life story has temporal structure. I feel consciously connected to Owen the boy born in 1949. I am the same person as him in one sense but not in other senses. I am physically continuous with that baby, and by virtue of being uniquely connected to and continuous with his body and brain and life, I possess the same unfolding autobiographical memory and the same personal stream of consciousness.

The second generalizable point about my self-description is that it conveys a certain way it feels to be me, which is not in every case explicitly stated (nor, perhaps, can it be), but that is sensed by both you and me. What I am able to say about my life comes with emotional coloration, even if these emotions cannot always be fully articulated.

William James described the stream of consciousness this way: Every state of consciousness is part of a unique personal consciousness. States of consciousness constantly change. At any given time, the stream attends to and is interested in some parts of the world and not in others. Consciousness is sensibly continuous and has an unconscious or semiconscious fringe.

The idea of the fringe is revealed phenomenologically, by introspection, but is also required as an inference to explain why some conscious experiences have the unique subjective psychic tone for me that they do. For instance, I know a sometimes mean-spirited Catholic priest, and when I am around him I get the heebie-jeebies in a way I don't when I am around other mean-spirited people. Why? The best explanation is that this priest reminds me of similar priests I knew as a boy, and arouses whatever bad associations of both fear and hypocrisy I still carry about mean-spirited priests. My stream is accumulative, it carries all manner of emotional coloration, it flows with a surround uniquely determined by my past experiences, some long forgotten or possibly never noticed, but nonetheless absorbed.

Still, there are senses in which I, Owen, the man of 52, am not the same person as Owen, the infant born in 1949. Owen the baby was small, Owen the man is large. Owen the baby was uneducated, childless, never married or divorced, had no siblings, had not lost his parents or a brother and a sister. I can't sensibly give a hoot about how the life of Owen the boy goes. It makes no sense to care about how it goes, it is over and done. I care a lot, however, about how his life *went*, since I am continuous with that person who once was. Who I am is in some respects importantly constituted by how his life went, how he was, who he was. His life is a proper part of mine. I try to understand his life because it tells me some about who I am. What I care mostly about is where things go for me from here. Again, the reason is simple: My future is open in ways that my past, especially my distant past, is not.

In the Monarch Notes version of my life I had to allow for a third salient feature of self-narration. My memory of my early childhood is now pretty dim, save for some emotionally important events. I suspect, although I can't prove it, that when I was five or six years old I could report with considerable accuracy how the previous days, weeks, and months had gone. My confidence that this is so is not due to any present memories but to the ample evidence offered by child psychologists that five- or six-year-olds can normally do this. From where I sit, these memories are faded or lost. But the child psychologists teach us that if you caught me at the right time, my recollection would have been pretty good. Not so, however, when I was one, two, or even three.[7]

Regarding the present, I am vividly aware of now sitting in my sister Nancy's kitchen overlooking the water on the Maine coast as I write these words. I feel the cool breeze from the water coming through the windows, and I hear the kids playing at the computer. What is sometimes called "the specious present" is vivid. Years from now I may remember sitting here, but if I do it will not be because it is so memorable in itself. It will be because I have put down the memory in words, fixed it as memorable, and joined it with strong feelings I have about my sister, Nancy, my nieces and nephews, and writing

[7]There are four events I seem to remember fairly vividly that all occurred between the ages of one and one and a half. To get to San Juan we flew from LaGuardia airport in New York to Miami on a large prop plane. I remember the sight and the noise of the plane on the tarmac, being held by Mom, but nonetheless frightened. On the boat to Miami, I remember Dad showing me flying fish and I remember acting like a maniac in the dining room (and getting into a fair amount of trouble). Sometime after we arrived in Puerto Rico I remember seeing several rats near the hedges. I was frightened. The fact that I remember these things, but little else from the first year in Puerto Rico, fits nicely with child psychology, which teaches that early childhood memory is poor, unless the remembered experience was emotionally very salient at the time of its occurrence.

about things I care about. In two days, when I am home again in North Carolina meditating before my Buddha altar with sandalwood incense burning, I will be in a different "specious present." The memory of the cool Maine breeze, Nancy's house, and the kids may well come up, but if it does it will be less vivid than the experience itself. It is a simple truth about most experiences that they lose vivacity over time. Often they are simply forgotten.

In any case, the story I told above consists of memories, surmises, and interpretation. But I haven't told you about just any old things that happened. I have told you about matters I deem important and self-revealing. The self involves more than just being or revealing a continuous person. It is an abstraction formed from the set of experiences, events, patterns, and processes to which we and others assign significance. This is the fourth generalizable feature of my brief self-narrative.

This assignment of significance is not random but depends on what Derek Parfit calls "psychological connectedness." I am the same person who ate *gallo pinto* (rice and beans) many days over the last two summers in Costa Rica. I didn't mention that in my abridged story because knowing about last summer's diet is relatively unimportant to either my own or to your fix on who I am. What we both care about are such things as my personal passions; important roles I fill such as being a father, spouse, or uncle; important falls such as result from an addiction or grave misjudgment; abiding traits of character; and significant achievements and goals. We care about these things because they reveal more about who we are than do more mundane things. Personal identity is grounded in continuity, but selfhood is grounded in psychological connectedness, in the connections to experiences, feelings, traits of character, and actions that we pick out of the complete continuous stream as most important, as most identity constitutive. Life's meaning accrues insofar as we express, experience, actualize a life that involves being a good person, as well as doing and experiencing good things, things that are important.

There is a fifth feature of my narrative worth marking—again because it is generalizable. I have referred to a time when I acted "out of character," a period when I wasn't myself. Neither our characters, our personalities, nor our agentic powers are so stable or resilient that once formed, we don't sometimes make mistakes, step off the track or even lose our way. It is a normative ideal that once the self takes shape and has form, we develop it and steady it, keeping it on track. But natural selves are imperfect. Biochemical changes in the "weather within" can bring a person low, making a happy person depressed. Even in good health, we are not always competent or wise: We can make decisions that are unwise or self-destructive. And the world can, for all manner of impersonal reasons, fail to cooperate with the aspirations, hopes, plans, and projects of a self along the way. It makes sense to have high ideals for ourselves. But again we must recognize that, as animals who are born of flesh, develop in idiosyncratic social environments, and face all manner of unpredictable contingency in living our lives, living well and maintaining our formed characters is a formidable task. But for that reason it is also a noble ideal, one we all have reason to aim at. Our integrity and the meaning of our lives depend on how we hold up as the person we are in the face of challenges.

No "Arch-Ego" Needed

I've been trying to say what a natural self, the only kind of self there is, is like. It will be useful to remind the reader of the sort of view that the naturalized picture of the self claims to replace, indeed, the one it claims is unwarranted and unnecessary.

The belief that humans must possess an indivisible soul, what James's called an arch-ego, was due in large measure to arguments of Thomas Reid and Joseph Butler to the effect that memory and the like, despite being evidence for the self, could not conceivably be what made up the self, could not be what the self is. Their argument was this: Consciousness of personal identity *presupposes* personal identity, and therefore cannot tell us what personal identity is or consists in. Psychological continuity or memory is *evidence* that I am the same individual person, but they are hardly what make me the same person. Squirrel footprints are evidence that a squirrel has been here, but the footprints are not what constitute the squirrel.

The best response to this objection—essentially it charges that the reasoning that seeks to explain personal identity in terms of the evidence provided by conscious memory is circular or question begging—involves a five-step response. First, we acknowledge that conscious memory continuity and the powers of agency displayed by creatures possessed of such continuity are offered as a hypothesis for what most people mean when they say or think that they are possessed of personal identity. Second, we claim that the accumulative stream of embodied experiences and actions best accounts for all there is to the ordinary sense of personal identity. Third, we claim not to presuppose a strong and determinate conception of what identity consists in. In particular, we deny that we do (or that our opponents should) presuppose that personal identity consists in an all-or-none condition such that a person is identical over time only if she possesses all and only the same features over time. Fourth, and instead, we propose that the sort of personal identity we sought to explain all along is a more-or-less phenomenon. It is a matter of degree. Individual humans by virtue of being possessed of the right sort of bodily and psychological continuity and connectedness possess personal identity, they are the same person over time, by virtue of embodying and realizing their own and only their own life. Fifth, and finally, we acknowledge that typically evidence for something is insufficient to say what the thing it is evidence for consists in or is. But in some cases, the evidence is sufficient.

The fact that the water in the well contains, upon chemical analysis, dangerous traces of arsenic is evidence that there is arsenic in the water. Furthermore, arsenic's being arsenic is all there is to there being arsenic. In such cases, the evidence is sufficient to tell us what the thing in question is. Psychological continuity might have been "mere" evidence for a self, for a person, especially if personal identity was characterized in advance as something all-or-none. If identity were necessarily permanent and immutable it could not, in principle, be something that could be captured, as what it is, by anything mutable, by anything which is merely continuous, even strongly so. But it is that view, one that seeks to account for what it is that makes some individual exactly the same person over time, that begs the question. There are no persons who fit the description of being exactly the same over time. The original question is ill-posed. And, as such, it disallows, in advance, all answers except for the metaphysically extravagant one it seeks—the one that says that it is possession of a permanent self or a soul that makes each individual exactly the same person over time.

The alternative view asks us to posit an abiding and immutable soul, James's "arch-ego." But in addition to being unwarranted by the facts—by the features of persons that make them more or less, but not exactly the same, over time—consider what an "arch-ego" would be like.

First, it looks as if it would need to be a completely structural feature of persons. If it held, say, some basic temperamental traits or the germs of one's personality, then it would seem to be made of features that change, develop, and occasionally disappear. So, second, to avoid this unsavory implication (unsavory from the perspective of the soulophile) it—the soul, the self, the arch-ego—will need to be impersonal.

But these two related features, the first, the need for the self to be purely structural, the second, that it be completely impersonal, cause a deep set of problems for anyone who would advocate the view that we are possessed of such a self. With regard to personal identity, if what makes me the same person is something structural and impersonal, then it is impossible to see how any self is to be distinguished from any other, at least when it comes to what allegedly makes that person who she is. Remember, the stream of consciousness, my experiences and actions, my character, my personality are all changeable, part of the flux, so these cannot make me, me. What makes me, me is my soul now understood to be structural and impersonal. But such a thing hardly seems like the right sort of thing to make me, me. Furthermore, insofar as my soul, my arch-ego, my essence is impersonal, and thus contains none of my contingent and unique characteristics, then what is it that makes my arch-ego mine, and what makes my possession of it my essence? According to the view on offer where the arch-ego is structural and impersonal, everyone might plausibly be said to possess the very same impersonal soul. And if so, nothing about the soul can distinguish any individual from any other.

Consider the related matter of personal immortality, which I claim is part of the motivation to ensoul us. If when I die and go to heaven it is only my immutable soul that goes, then I don't gain personal immortality in any sense. There is nothing of "me" in my arch-ego. If, on the other hand, my soul passes into the hereafter with "me" attached, then I do achieve personal immortality—this person who is me gets to heaven, after all. But then the soul or arch-ego doesn't seem to be doing any work save as serving as the container in which my historical self—the only self I care about—is wrapped. And it serves the same function for every other actual person. If God puts a soul into me that contains the germs of my character or personality, then my soul, contrary to what it is supposed to be, contains mutable components. Furthermore, if God endows me with the germs of a personality when he endows me with my soul, and if, as seems plausible, my primal personality gives me good, bad, or neutral equipment for leading a successful life, it is hard to see how God can be understood as being fair.

The soulophile faces a dilemma from which there is no escape: Either what makes me, me is purely structural and impersonal, in which case it explains absolutely nothing about what makes me the unique person I am; or the soulophile allows my arch-ego to be somehow constituted by some (or all) of the features of my character and my life that make me, me, in which case what makes me, me is mutable, and so on. So either the soulophile grants me an impersonal soul that makes me, me according to her lights, but says nothing about what makes me the actual person I am, or she concedes that there is nothing about me that is immutable, indivisible, permanent, and so on. Pick your horn.

All of these puzzles are avoided by the naturalist's picture. Thanks in large measure to a particular act of sexual intercourse by my parents, a conceptus was formed. I (or the set of cells that were to become me) was launched with a unique genetic blueprint that contained instructions for a particular human being, a potential person, quite possibly, with certain natural temperamental and characterological dispositions (there is, for example, some pretty good evidence that some infants are born quieter, more reserved, even shyer, than others). Thanks to the organic integrity I possess, the psychological equipment housed in my body that awaited development in the world, thanks to the particular family and larger social environment I was born into, and within whose structures I took form, I emerged.

Once I emerged, I quickly picked up on how to express my self, to gain some of the things I wanted, to cooperate, and eventually to take myself out into the world and make something of myself.

The naturalistic picture explains everything in need of explanation. There is simply no need for a nonnatural or supernatural posit to explain the existence of the sort of selves that abound in this world.

Humans, Persons, and Selves

John Locke famously distinguished between a human and a person. A human is a continuous spatio-temporal animal. A human, a "man," or better a "mere man" as Locke called such a thing, could be thoroughly amnesiac about its life. So long as we can plot a man's spatio-temporal trajectory from birth to death, even a completely amnesiac man, we have the same man. In this sense, a "man" has identity conditions like a plant. A plant is the same plant so long as we can plot its life in spatial and temporal terms. The identity of a plant rests not at all on its possessing consciousness of its life.

A "person" differs from a "man." A person, Locke tells us, is a normative, specifically a "forensic" concept. A person possesses both the bodily continuity of a "man" and psychological continuity or memory. It is conscious memory that makes for a person.

Locke is concerned with the ways we use the concept of "person" in moral and legal practice. We hold nonamnesiac "persons" accountable for their actions in ways we don't hold amnesiac "mere men" accountable.[8] Normal members of the species *Homo sapiens* possess memory and are thus persons. Since an embryo has no memory, it is better to say that it possesses the potential to develop capacities for memory, and that barring some misfortune, it will normally develop to possess memory.

The concept of a mere man is something of a fiction, at least insofar as it relates to moral accountability. Persons are ubiquitous among members of the species *Homo sapiens*, whereas mere men, that is, men

---

[8]I'm going to avoid discussing a genuine problem with Locke's view, namely, that forgetfulness of some criminal or immoral act might be exonerating. This is an important issue, but tangential to my present purposes.

who never become persons, are relatively rare. The profoundly retarded may qualify. To be sure, some individuals do go from being mere men *in utero* to being persons to being mere men again. Patients suffering from Korsakoff's disease (which results from destruction of the mammillary bodies in the brain due to alcoholism) or Alzheimer's disease often lose their memories to the point that we rightly feel there is no longer "anyone home." Such individuals, formerly persons, are then mere men.

In human embryos, Mother Nature creates mere men who then in the normal course of development become persons. What she does not create are selves in the robust sense of the term. She provides the basic apparatus on which selves can be constructed, the blueprint for a continuous body and brain, and for memory. But rich and robust selves are largely the product of social interaction.

It is worth noting that Locke's distinction between a "mere man" and a person corresponds roughly to Damasio's distinctions among creatures possessed of a proto self (they are unconscious but can get around) and those with both core consciousness and autobiographical consciousness. The latter are persons, according to Locke. A person has a stream of consciousness and she can stop and hold in mind certain events in the stream. What we might call a "mere person" doesn't make use of her capacities to pick out anything as more important than anything else. If we imagine such a lost soul speaking about her life, she would provide a one-thing-after another litany of what she remembers. Such a mere person possesses Damasio's core self, as well as an utterly undisciplined autobiographical consciousness. Actual persons, normal ones at any rate, fix on certain experiences of external states of affairs and on certain autobiographical experiences as worth remembering, reporting, and scrutinizing. When actual persons focus on autobiography, they are focusing on the self. The self is an abstraction from the story of a person's life that isolates and magnifies the experiences, traits, and aspirations that are assigned importance.[9]

[9]The attentive reader might have noticed that for all I have said that a chimp, bonobo, or great ape can't be a human (wrong species), I did not say it could not be a person, and might even be, or have, a self. This is fine with me. Although I remain agnostic on the matter.

My True Self

Selves are created in two main ways—by the participation of persons with the natural and social world, and by cognitive abstraction, from among the entire inventory of first-person memories, of the events and patterns that are most revealing, explanatory, and predictive. Maturation, living in the world, being socialized into a way of life, and acquiring certain ways of thinking are needed before one can start to build the complex set of representations that, taken together, come to constitute one's self.

In saying that the self is an abstraction from personhood,[10] I mean, at first pass, this: Things are done by us and to us, and things happen around us. We are disposed to abstract from this multifarious set of doings and happenings those that we think reveal a pattern that is predictive of our personal trajectory and that capture the ways we normally feel, think, and act, what we know about, and, more importantly, what we care about. We abstract for a purpose. The self is an abstraction designed to do, in interpersonal and intrapersonal commerce, the work of explanation, prediction, and control.

Although specific answers to questions of identity vary, these general purposes do not. In some places and times, one's being a father will be centrally self-constituting, in other times and places it will be incidental. Paternity might matter, but being a father who is involved in loving, caring for, and raising one's children might not. What specifically will be highlighted in the narrative will vary, but the general goal of self-narration does not. Its function is to situate those aspects of oneself for oneself and others in order to engage in the projects of self-knowledge, self-explanation, self-prediction, and self-control, and to assist others in so doing.

Whatever else we mean by the "self," it is constituted only by a subset of all the memories a person, assuming she had perfect memo-

[10]One might think that saying that the self is an abstraction is in tension with my naturalism. It isn't. The thought that it might be would rest on the premise that a naturalist can only admit natural, that is, physical things into her universe. But every sane naturalist will allow abstract concepts (which may well be explicable in terms of shared ideas that are realized somewhat differently in different brains), so long as these abstractions are rooted in the natural world.

ry, could hold in mind. I, for example, could tell you about all the things I remember doing and happening since I woke up today, but most of this would reveal nothing about *me.* I want to remember and you want to hear about *me,* about who I am, about what matters and what is important. The self, the unique person I am, is occluded from view in a complete and unedited memory report of all that has happened to me and all I have done.

The self is an abstraction that is commonly captured in a narrative structure. Self-narration has temporal order, it marks certain events as more or less significant, and it involves theorizing. By this I simply mean, in the first instance, that our self-conception is typically filled with hypotheses about cause and effect, and these do not involve memories but surmises: "My father was stern and this led me to be timid." Often, these interpretations are framed in terms of culturally available psychological models—psychoanalytical, behavioristic, psychobiological models, for example. The manifest image is invariably a central, dominating influence in how we frame our story.

Finally, the self is an abstraction that contains trait-posits—character traits, virtues, and vices that can form a basis for reliable explanation and prediction about my feelings, thoughts, and actions. When I posit character traits I am not viewing these traits. I am surmising that I "have" them based on certain patterns in how I think and behave (or think that I think and behave).

So far I've been running together the distinction between the self and the self-narrative, where the self is the abstract entity about and around which the narrative clusters. Some think the self is *all* narrative. This idea is not crazy, but it is somewhat misleading and thus unhelpful. Think of the differences in description from the first-person and third-person points of view. It is commonplace that others sometimes see me truly in ways I don't see myself. A friend says that "You are shy," but I don't see myself this way. We talk. Our talking about the way I really am might make it seem as if there is some third thing separate from our two ways of talking that is my self, "my true self." And there is. This third thing, however, is not a thing in the strict sense, or, at least, in the usual sense. What there is—indeed what there must be—is a set of patterns we are trying to settle on a description of, a helpful, explanatory, and predictive description. It seems plausible to say that the pattern in question either was or wasn't there before either I or my friend saw it. And thus that it was a pattern correctly or incorrectly ascribable to me before either of us discerned it. At the point of disagreement there are two discrepant narratives. My self as seen by me is non-shy. But I am shy as seen by my friend. One description is undoubtedly better than the other at revealing the patterns in question. One way to make this point is to note that my self is constituted on most any plausible view of behavioral patterns, as well as patterns of feeling and thinking that might not yet have been revealed explicitly in any narrative but that are there, there to be discerned by an acute eye. The main point is that I was either shy or non-shy before anyone said so.

This suggests that it is both credible and legitimate to posit a real or true self that is not completely captured in narrative. One's true or real self is constituted by the description of a self that would be captured by an omniscient observer of who that person is and is like. The omniscient observer would need, besides knowing all possible facts about that person, to possess the right theory in which to frame what makes a particular person who she is. This observer might well think that the person is mistaken in certain of her own surmises, that she is more or less psychologically connected than she thinks to certain experiences, projects, events, and actions.

Now the soulophile might see an opening here, and even a fellow naturalist might think I am playing with fire in suggesting that we posit a real or true self. Never fear. The soulophile will find no consolation in the true self, nor should my fellow naturalists have anxiety attacks. The real self is not an extra ingredient in you or a further fact about you, as the soul is. It is more like the total amount of energy or matter in the universe: An omniscient physicist, if there was one, could state the precise amount, even though, to us, it is unknown. The real or true self is there. There is a determinate set of patterns that embody who I really am and what I am like, and it would be seen for what it is by an omniscient psychologist if there was such a thing.

I have said that I favor thinking of the self as an abstract theoretical entity. I feel I've thwarted the soulophile who thinks in saying this I conceded something very much like the existence of the soul, and I hope I've calmed any of my fellow naturalists who entertained similar suspicions. But now I see the postmodernist I described earlier coming over the horizon, licking his lips as he contemplates devouring me in his next meal of wayward analytic philosophers, saying "Yup, there you've admitted it, the self is just a fiction."

But I did not say that, or even imply it. What I said was that the self is an abstract theoretical entity in the same way that force, mass, and energy are abstract theoretical entities. Einstein taught us that the total amount of energy in the universe is equivalent to the total amount of matter in the universe multiplied by the speed of light. Matter and energy are in an important sense interchangeable: One can be converted to the other. No one knew this before Einstein, but it was true before Einstein said so. Energy is an abstract idea but it also exists in the real world. Likewise, the self is both a real entity and an abstract idea intended to pick out the most important features of an individual's life. The self-concept a person holds, the ways in which she represents herself, and the way others see her are designed to capture her self, but because of our fallible cognitive capacities, not all the features of the self are seen clearly. We at best approximate capturing the self in our first- or third-person thoughts and stories.

History is another useful analogy. Surely, human history exists. It is not a fiction. Every historian tries to pick out what is most important, explanatorily relevant, and predictive. An omniscient historian could do so. But no actual historian is omniscient. The history an omniscient historian would describe would be true and complete. Actual, very smart historians cannot do this. Not because the history and the relevant patterns aren't there, but because real historians are cognitively limited.

Just as it makes sense to distinguish history—even of some relatively brief historical segment—which is the object that real-world historians seek, but invariably fail, to penetrate fully in their rich narratives, so too is it advisable to distinguish between the self and the narratives that aim, but invariably fail, to capture the self. The postmodernist can now, please, leave the premises, without having earned the right to have me for a snack.

At any given time I hold a self-model or self-conception in my head. My self-model is a complex, dynamic, retrievable, and expandable dispositional structure of who I am. It gains in complexity and texture as I experience and do new things. Sometimes, without doing a conscious self-inventory, I find myself thinking or saying totally new things about my self. Once consciously articulated and thus noticed, these novel observations are incorporated into my self-model. The self-model contains unconscious, preconscious, and conscious components. Sometimes, new self-observations bubble up as newly thought or expressed parts of the self-model already in our head. Other times they emerge out of the preexisting model's interactions with new experiences. Often there will be no certain way to know whether we are expressing something already dimly or inchoately known or making a totally new discovery. Usually, it won't matter.

The Narrative Self, Reprise

Work on the narrative self is currently a hot topic in psychology, literary theory, and philosophy. Some think that the self is narrative all the way down. I deny that, since I believe in true selves, selves not captured, possibly not capturable by finite human minds in any story. One might think that I could still claim that the self-model, as opposed to the self, is narrative all-the-way-down.

And indeed I could claim that. It would not be inconsistent with anything I've said so far. But I won't say it because I think it is not true. The reason I think this would be the wrong way to go comes, first, from what I said about conceptual structure earlier. To possess an autobiographical self, one needs to possess concepts. But, partly out of respect to my friends the chimps, bonobos, and gorillas, I resisted requiring that all concepts be captured linguistically. Second, as for human selves, possessed of a self-model, much of what we find ourselves self-revealing is not explicit—never has been. I say or reveal certain things about who I am as situations call for certain kinds of speaking or revealing.

Third, most contemporary cognitive scientists think that Freud was at least partly right in attributing defense mechanisms by which we deflect or repress certain painful experiences. And we all believe in implicit or subliminal memory. Things we do not report, some we cannot report as remembered or known, were in fact remembered. Perhaps there was the atmosphere of one's family of origin, in one's childhood home, that was experienced for what it was—happy, busy, threatening, silencing, or distant. Since there was at the time nothing to contrast this atmosphere with, the feeling might have been absorbed, but possibly only in an unarticulated fashion. It now colors what James called the "fringe" of consciousness. Since the fringe partly constitutes the stream that constitutes me, it is part of my self. But if I do not or cannot access it, it is not part of my self-concept, my self-model.

If, with the aid of a savvy therapist or even on my own, I engage in severe and sustained self-scrutiny and I come to see, understand, and describe the atmosphere of my home, it will then, of course, become part of my narrative. But it is not clear why we would want to say—as a proponent of the self-as-narrative-all-the-way-down—that the atmosphere in the house in which I grew up was stored in narrative form before I was able to speak about it in linguistic terms. Furthermore, there is no guarantee that for every important causal contributor to who I am—even for certain aspects of my life stored or represented in fringelike form in my self-model—that I will be able to see it, capture it, and articulate it, even under the best of circumstances. It will be there in the fringe, nonetheless, in what James sometimes calls "the halo." Having said that our self-model is not best thought of as narrative all-the-way-down—neither composed of narrative all-the-way-down nor completely expressible in narrative form—the self as expressed in narrative, the narrative self, plays an especially important role in human life.

Self-Representation

We represent our selves to answer certain questions of identity. Given that I am a person, what kind of person am I? Getting at identity in this sense requires more than detecting that one is the locus of a certain kind of psychological connectedness and is continuous with a single organism over time. It requires that one have representational resources and cognitive abilities to access and articulate the various states, traits, dispositions, and behavioral patterns that make one the person one is.

We may distinguish two different aims of self-representation. First, there is self-representing for the sake of self-understanding. This is the story we tell ourselves to understand who we are. The ideal here is convergence between self-representation and an accurate story of who we in fact are. Second, there is self-representing for public dissemination, whose aim is to underwrite successful social interaction. The two are closely connected. Others, watching for consistency between what we say we are and how we behave, provide various kinds of subtle and not-so-subtle feedback about the accuracy of our self-descriptions. If others see us as providing a far-fetched account of the self, they will let us know, sometimes by having nothing further to do with us.

Self-represented identity, when it gets things right, has one's true nature (or some aspect of it) as its cognitive object. But representing one's self does not leave things unchanged. The activity of self-representation is itself an experience that influences the self. This is true in two senses. First, considered purely as a cognitive activity, representing one's self involves the activation of certain mental representations and cognitive structures. This activity inevitably realigns and recasts the representations and structures already in place. Second, the self as represented has motivational bearing and behavioral effects. The function of placing one's self-conception onto the motivational circuits involves certain gains in ongoing conscious control and in the fine-tuning of action. In other words, everyone wants to be a certain way; self-representation calls one's attention to discrepancies between what one is and what one wants to be, between where one is and where one wants to get to.

Sometimes, especially in cases of severe self-deception, the self projected for both public and first-person consumption may be strangely and transparently out of kilter with what the agent is like. In such cases, the self as represented is linked with the activity of self-representation but with little else in the person's psychological or behavioral economy. Nonetheless, such misguided self-representation helps constitute the misguided person's identity. It affects who she is.

My Buddy Billy Fletcher

I first wrote about Billy Fletcher in my book *Self-Expressions.* My students see the story as an amusing anecdote about their still absent-minded professor. But Michael Gazzaniga, one of the founders of the field of cognitive neuroscience, has written about it not simply because it is amusing. It reveals that our narratives can contain falsehoods that are nonetheless self-constituting.

When I was 26, I returned to New York from Cambridge, Massachusetts, to spend Thanksgiving with my parents and my five siblings. Over dinner one of my sisters asked my parents a question about the anxieties of being young parents of young children. My mother reported that one of her most worrisome times was when she, my father, my sister, and I returned from our four-year stint in Puerto Rico so that my father could take a new job in the States, and so that I could start first grade. Her worry was that at age five, without the experience of kindergarten, and coming from Puerto Rico where I "had no playmates," I might have trouble adjusting to school.

I was incredulous and reminded my mother of my deep and important friendship with Billy Fletcher, whom I played with "all the time." My parents both looked at me as if I were crazy. Mom explained that Billy Fletcher was the son of a business relation of my father's who had visited Puerto Rico for a few days and had spent exactly *one* afternoon at our home, during which time I had indeed played with him.

Several distinguished child psychologists have reassured me that childhood confabulation of this sort is not uncommon. There were lots of photos taken that momentous afternoon and no doubt looking at them helped fill out my imaginary friendship with Billy. The fact remains that had my mother never told me the true story I would have believed my version. I might still be embellishing it with new memories.

I want to make three points about this tale. First, what I remembered about my early years was, in one significant respect, simply false. Second, the confabulation is best explained as involving a germ of truth planted in an autosuggestible host. Quite possibly I was lonely and was fertile ground for the elaboration of my notional friendship. Third, my extended fantasy of playing with Billy was identity constitutive in two respects. I was *really* Billy's friend and playmate, *and* being his friend and playmate, even though only in imagination, prepared me for getting along with other children when first grade began.

How Fictional Is the Narrative Self?

So whom does my narrative self play for? The answer should be pretty clear by now. It plays for third parties, and it plays for one's self.

The "me" for which my narrative self plays is the whole embodied information-processing system from which the narrative self emerges and for which it plays a crucial role. Often I express and grasp snatches of my self and utilize these representations in monitoring and guiding my life. This way of thinking about our reflexive powers makes them fairly unmysterious. I am a system constituted in part by a certain conception of self. Sometimes I hold that conception in view as I think about things. I am more than my narrative self and so can comprehend it, utilize it, and, in concert with outside forces, adjust it. What am I? This organism. This thinking thing. This person.

Daniel Dennett insists that the self that is the center of narrative gravity is merely a useful fiction. "Centers of gravity" are "fictional objects . . . [that] have only the properties that the theory that constitutes them endowed them with." What might this mean? If our self-model is a fiction, then the postmodernist's position is won. Dennett is no postmodernist, so I need to save him from his own words. Interpreted charitably, Dennett is best read as pointing out the ways in which self-representation involves fictional aspects. No one claims that the story conforms to the sort of representation one expects of, say, an ornithologist describing the life of some bird he has followed from hatching to becoming a hawk's supper. Nor is it like the record the video camera at an ATM machine provides of all its "visual experiences." Every sensible person acknowledges that self-representation is interpretive, fallible, and to some extent self-serving. That said, it does not follow from the premise that self-representation contains fictional aspects that it is fictional through and through. But Dennett seems to say this. After saying that the self is a fictional object he writes, "fictional objects . . . [that] have only the properties that the theory that constitutes them endowed them with."

In the summer of 2000, I read three autobiographies: *The Story of My Life* by Helen Keller, Benjamin Franklin's *Autobiography*, and Linda Brent's *Incidents in the Life of a Slave Girl.* I assume that all three autobiographies involve interpretation, contain mistakes born of forgetfulness and poor surmises, and contain some self-serving spin. (Interestingly, Ben Franklin, who trained as a printer, calls poor judgments "errata" in his autobiography.) But I don't think these self-narratives are remotely like Dorothy's story in *The Wizard of Oz,* or like Hamlet's or King Lear's in Shakespeare's plays. I don't think Dennett thinks so either. Let me explain what he might sensibly mean in asking us to think of the self as a fiction.

The idea that the self is a fiction is in part a way of saying that it is a construction. Mother Nature does not start us off with a full-blooded self. She starts us off caring about homeostasis and equipped to distinguish "me" from "not me." She does not wire in a personality or an identity. Identity is the joint production of many sources, including one's own developing self and the ever-evolving story one uses to locate who one is. The self might be construed as a fiction because it is constructed and because the narrative with which we construct and express it is open-ended.

The idea that the self is a fiction also captures a second feature of identity. The self is subject to constant revision in two senses. Not only do new things I do change the ongoing story, but the past is sometimes reconstructed or filled out using hindsight. Sometimes these reconstructions are self-serving ("I never really loved her anyway"), but other times they involve rendering various indeterminacies determinate, revising certain hypotheses about what we are like in light of new evidence, and answering questions that hadn't arisen before. For example, wondering in one's forties why one cares so deeply about a certain worthless thing, one reconstructs some story of one's distant past to explain it to oneself, and possibly to others. Most such reconstructions are uncorroborated; some are uncorroboratable. But there are constraints, and we and others watch for reconstructions that make sense.

So the self is said to be fictional because it is a construction and because it involves all manner of revisitation to past indeterminacies and reconstruction *post facto.* Dennett has us imagine John Updike making the Rabbit trilogy into a quartet not by writing about a still older Rabbit Angstrom (which, in fact, Updike has now done) but by creating the story of a very young Rabbit Angstrom. The extant trilogy constrained what Updike could say about the Rabbit who existed before *Rabbit Run,* as well as what it could make sense to say about an older Rabbit—although for a host of reasons what can be said about an older Rabbit is much more open. But the vast indeterminacy of Rabbit's former life creates numerous credible ways in which the story of the earlier years could be told. For certain parts of our lives, we have similar degrees of freedom in how we tell the story of our selves.

There is a third way in which the self may be considered a fiction. Fashions in storytelling change: The kinds of heroes, villains, and dilemmas we find most compelling vary across cultures and generations. In the same way, a life is satisfying from the inside and respected from the outside when its central themes are built around worthy aims and values, but these values vary temporally, culturally, and subculturally. Different kinds of narratives fly at different times and in different places. The book-buying and interpersonal markets create selection pressures favoring certain narratives and disfavoring others. Being a computer nerd was a somewhat different matter in the 1970s from what it was in the 1990s, especially when the latter version drove a Lexus and his company had just gone public.

This brings me to a fourth way in which the self might be said to trade in fiction rather than fact. The self that is the center of narrative gravity is constructed not only out of real-life materials; it is also organized around a set of aims, ideals, and aspirations of the self. Since these have not yet been realized, they are fictions, albeit useful ones. The concept of the self as fiction, in short, might seem right for these four reasons:

- It is an open-ended construction.
- It is filled with vast indeterminate spaces, and a host of tentative hypotheses about what I am like, that can be filled out and revised *post facto.*
- It is pinned on culturally relative themes.
- It expresses ideals of what one wishes to be but is not yet.

But there is a crucial respect in which the self is not fictional. The author of a piece of fiction has much more freedom in creating her characters than we have in spinning the tale of our selves. Self-construction is constrained by reality: the things we have done, what we have been through as embodied beings, and the characteristic dispositions we reveal in our personal and social lives. Others will catch us if we take our stories too far afield, and we may also catch ourselves. Social selection pressures keep the story that one reveals to one's self and to others in reasonable harmony with the way one is living one's life. To ignore these pressures is to court madness.

The self can be a construct or model, a "center of narrative gravity," a way of self-representing, without being a fiction. Biographies and autobiographies are constructs, but if they are good, they are nonfictional or only semifictional.

If we were to interpret Dennett as holding the implausibly strong postmodernist thesis that all self-narration is fiction, then he like postmodernists should favor dissolving the distinction between fiction, biography, and autobiography. Some students and scholars at eminent research universities, including my own, have in fact swallowed such a leveling of genres. But good sense still prevails among the people in charge of library catalogs and bookstores. Books are organized in such places in accordance with books of like genre for more than convenience. There is a matter of principle involved. Biographies and autobiographies are not works of fiction.

Some people are, of course, massively self-deceived, including some who write autobiographies. But self-deception can exist only if selves are not totally fictive—only if there are some constraints on what may go into our self-narrative. Self-deceived individuals, possibly totally unconsciously, keep certain facts from entering the narrative. Some alcoholics know that they have a problem but try to keep their drinking a secret from others. They deceive others but not themselves. Other alcoholics display their alcoholism publicly but develop immunities to comprehending social feedback intended to challenge their self-conception that they have no drinking problem. They deceive themselves but not others. Self-representation, even massively deceived self-representation, is causally efficacious—it causes the person to say wildly false things about himself.

This would not matter if the self were entirely fictional, but it obviously does matter.

A person's self-model is one of the many models contained in his mind/brain. Once acquired and in operation, this model is part of the recurrent brain network causally responsible for the life a person lives and how he thinks and feels. This is true whether a person's model of his self is in touch or out of touch with who he is and what he is really like, and it is true whether a person's self-model contains worthwhile or worthless aims and ideals. Even though the self is not *the* control center of the mind—the mind may well have no single control center—in setting out our plans, aspirations, and ideals, our self-model plays an important causal role in a person's overall psychological economy.

The self-model changes, evolves, and grows. The fact that my conception of my self is a model housed in my brain explains the first-person feel I have for my self (but not for your self), and it explains my special concern that my story go as I plan. If things go awry, if my plans don't materialize, if great pain befalls me, it will happen to this subject of experience, to the individual wrapped in this particular narrative. It is not surprising that I care so deeply that the story go the way I intend.

To be a whole person is to have narrative connectedness over time, caused in some fair measure by the authorial work of the agent. When we dream, fantasize, or dissociate we may to varying degrees lose this connectedness by deliberately or unconsciously ceasing or perhaps revising our authorial work. But these are exceptions—respites from the lifelong work of constructing the self.

Lost Souls and Trauma

The senses of identity, direction, agency, our projects and plans are grounded in the memorable connections of the embodied stream, in an assessment of who one is, what one is like, and what one aims to become and accomplish. When we assess who we are and how we are doing, we are examining our lives retrospectively. When we formulate new plans and projects, or see how far we are on the way to meeting certain set goals, including ones for self-improvement, we are projecting ourselves and our lives down the road. We project into the future a (hopefully) accurate sense of who we are now, where we have been, what we have done, and what we are like. Plans and projects need to be constrained by a realistic assessment of who we are and what sort of future might be in the cards for us.

It follows that certain extreme defects of consciousness should have effects on selfhood. Sometimes mental defects are so great that we no longer believe that Jane now is the same person she was before. Certain sorts of brain damage, for instance, eliminate the conscious stream altogether. Final-stage Alzheimer's patients are like this. An intact brain stem will keep one alive, but the destruction of brain tissue, especially of the cerebral cortex, destroys the body parts needed to house and access the autobiographical self.

Even without severe brain damage, the loss of the sense of identity, of personal sameness, can undermine our selfhood. The famous psychiatrist Erik Erikson was the first to write of what is now known as "post-traumatic stress disorder."

> The term "identity crisis" was first used, if I remember correctly, for a specific clinical purpose in the Mt. Zion Veterans' Rehabilitation Clinic during the Second World War. . . . Most of our patients, so we concluded at that time, had neither been "shell-shocked" nor become malingerers, but had through the exigencies of war lost a sense of personal sameness and historical continuity. They were impaired in that central control over themselves for which, in the psychoanalytic scheme, only the "inner agency" of the ego could be held responsible. Therefore I spoke of loss of "ego identity."

Individuals in identity crises are persons in that they normally experience themselves as the locus of a set of subjectively linked events, as a conduit in which a certain bland sameness and continuity subsist. But nothing is sensed or perceived as more important than anything else. Individuals with what is now dubbed "post-traumatic stress disorder" may suffer a detachment or estrangement that makes them akin to what I earlier called "mere persons." Things happen to them and in them, but nothing stands out as having any more importance than anything else has. What they lack, and what horrifies us and immobilizes them, is any sense of coherent and authoritative me-ness, of "personal sameness"—any sense that these subjectively linked events occurring to and in them constitute a bona fide person, a self, a life.

Erikson asks what "identity feels like when you become aware of the fact that you undoubtedly *have* one," and he answers—sounding more like James's disciple than Freud's—that it consists of "a *subjective sense* of an *invigorating sameness* and *continuity*." Without the sense of self, there is no coherent cognitive and motivational core from which the individual can generate purposes or find the energy to sustain them. There is a linkage between a firm sense of identity and the capacities to formulate goals and sustain effort. Emptied of "the subjective sense of an invigorating sameness and continuity," consciousness has no role to play in moving one's life in a certain direction.[11]

Individuals in identity crises have memories, but these simply don't feel like *their* memories, at least not in the usual way. In other sorts of cases, memories feel owned but are of such short duration that they can hardly be said to constitute a self. In *The Man Who Mistook His Wife for a Hat*, Oliver Sacks describes "The Lost Mariner," a patient named Jimmie. Jimmie lost his long-term memory to alcoholism. His impairment, Korsakoff's disease, the destruction of the mammillary bodies in the brain by drink, left him in a state in which "whatever was said or done to him was apt to be forgotten in a few seconds' time." Jimmie was "isolated in a single moment of being, with a moat or lacu-

[11]The naturalist will understand cases like Erikson's as involving changes in the brain, possibly mainly changes in brain chemistry, for example in levels of various neurotransmitters. The terrible experiences of war cause brain changes, which cause mental changes, which cause changes in thought, motivation, and behavior. I point this out just to remind the reader that the scientific image that regulates psychology assumes "supervenience" of the mental on the physical, that is, that any change at the level of experience is due to a change at the level of the brain. There may be plenty of brain changes that are not detected or detectable phenomenologically, even by disciplined introspection. Such changes at lower levels might well nonetheless have effects of how we feel overall, and on how we attend, deliberate, think, and act.

na of forgetting all round him. . . . He is a man without past (or future) stuck in a constantly changing, meaningless moment."[12]

Jimmie is conscious. He has short-term memory, some form of Damasio's core consciousness. No person has ever existed who was conscious but lacked short-term memory, this being necessary, and possibly sufficient, for holding an experience long enough to have it. But his short-term memory is *very* short. Jimmie's short-term memories are hardly sufficient to constitute a dribble, let alone a stream, of consciousness. His inner life seems choppy—Sacks describes it as "a sort of Humean [*pace* Hume] drivel, a mere succession of unrelated impressions and events." Indeed, when asked, Jimmie denies that he feels alive. He is neither miserable nor happy. His experiences don't connect in a way that constitutes a self that could be any way at all.

Amazingly, Jimmie discovered a way to reconstitute his identity. First in chapel and then in art, music, and gardening, Jimmie's attention was held. What sustained him and his activity in these contexts was not so much a stream of his own bound by contentful memories but a stream bound by affect, by mood, and by aesthetic, dramatic, and religious resonances. In these self-reconstituting moments it was almost as if Jimmie's conscious stream burst forth into a thick and rich flow that was Jimmie's own. When mood and affect rather than explicit memory reconstitute Jimmie's stream, he feels alive, acts alive, and seems happy. The reemergence of a personal stream coincides with the reemergence of a person, a bona fide self.[13]

Zazetsky, the brain-damaged soldier described by A. R. Luria in *The Man with the Shattered World*, works tirelessly to recover his self by producing on paper a narrative in which he appears as the main character. Zazetsky had two worlds shattered at once by a single bullet.

[12]Actually Jimmie has anterograde amnesia but not retrograde amnesia. That is, he has autobiographical memory for the time, prior to the end of the Second World War, before he lost his long-term memory, but having lost his long-term memory for postwar events, he cannot take his self forward.

[13]It is hard to say what sort of sense of self Jimmie possesses at these times. It is not the kind that can be put very well into words. His self at these times is full, or fuller than it normally is, but is not linguistically or conceptually structured. It might be almost fully composed of the "halo" or "penumbra" that for the rest of us normally constitutes the "fringe" of the stream.

The external world is disjointed, gappy, and strangely textured. Zazetsky's inner life is also a mess. His life lacks wholeness, meaning, and direction. He cannot find his self; he cannot relocate who he is. Zazetsky "fights on" to reconstitute his self by valiantly trying to uncover memories he still possesses but is out of touch with. Automatic writing, just letting his hand mark on paper what it wants to produce, results eventually in the production of a three-thousand-page autobiography.

Zazetsky produced, or better, he reproduces the objective story of who he is. The story of his schooling, his military career, his talents—chess, for example—can be retrieved, but not felt as truly his own. How he felt as a schoolboy, soldier, or chess player is strangely missing from his story. His autobiography is more akin to a 3,000-page résumé than a life story. Sadly, Zazetsky himself is impaired in ways that keep him, the very subject of the autobiography, from consciously reappropriating the self revealed in his tale. Still, he sees that "fighting on" to get his story out and bring it into view is his only hope for reentering his self. This is why he writes tirelessly. Identity is that important.

No one can grasp the whole narrative of his or her life at once. There is always more to our identity than what we can narratively convey. For one thing, we are forgetful. Zazetsky's problems are extreme, largely because his memory is so bad that he can barely keep in view snatches of narrative long enough to write about them. But his inability to grasp his identity all at once distinguishes him from no one. Furthermore, no matter how successful Zazetsky became at regaining his self, he would not capture his full identity. He cannot recapture his true self. No finite, fallible, defensive creature can succeed completely at doing so. In this sense, Zazetsky's quest is universal. But his deficiencies at recapture are so extreme that his plight is tragic.

Jimmie and Zazetsky have serious problems caused by specific brain injuries that have impaired their autobiographical memories. Other brain injuries result in personality changes that despite being peculiar and out of the ordinary do not impair the narrative self as badly. In *Descartes' Error,* Antonio Damasio recounts the tale of Phineas Gage. Gage was a railroad worker who in 1848 took a 13-pound rod,

3 feet 7 inches long, 1/4 inch in diameter, straight through his head. He was tamping an explosive charge with the rod when he inadvertently set off an explosion. The rod entered his left cheek and exited the top of his head, landing one hundred feet away "covered in blood and brains." Gage was stunned but never lost consciousness. Despite developing a bad infection and losing vision in his left eye, he soon returned to normal life, but as a very different person. Before the accident, Gage had been a sensible and temperate man, possessed of stable and reliable character. John Harlow, his doctor at the time, reported that "the balance between his intellectual faculty and his animal propensities has been destroyed." The new Gage was fitful, profane, intemperate, disrespectful, obstinate, impatient, and reckless. According to Harlow, "there was no sense of his future, no sign of foresight."

Gage died at 38 but his skull and the iron rod have been on display at the Museum at Harvard Medical School since 1861, after Harlow received permission from Gage's sister to have the body exhumed for scientific examination. Over a century later, Hannah Damasio and her colleagues set to work, using the latest in new computer and brain imaging technology, to try to reconstruct what parts of Gage's brain had been destroyed. It was clear from Gage's life that the areas that support language and motor function had been spared, as were most of the areas that support attention, memory, and the ability to perform calculations. The main damage involved massive insult to the left prefrontal cortex, significantly less to the right prefrontal cortex, in particular to areas beneath the exterior surface of cortex close to—perhaps on—the midline between the left and right hemispheres, in the ventromedial region of the frontal lobe. The damage to this area, what Damasio calls the "underbelly of the frontal lobe," is what "compromised his [Gage's] ability to plan for the future, to conduct himself according to the social rules he previously had learned, and to decide on the course of action that ultimately would be most advantageous to his survival."

Phineas Gage is different from Jimmie and Zazetsky in that his capacities for self-narration were largely spared. But there is a lot of explaining to do about why his personality is so different after the accident. There is much evidence about how Gage behaved after the trauma, but virtually none about how he described himself. We might imagine him explaining the change in himself in these foul-mouthed terms: "You'd be fucked up too if a steel rod knocked out your right eye, lifted you several feet off the ground, and spewed your blood and guts onto the ground a hundred feet away." Neither he nor Dr. Harlow would have been crazy if they thought Gage was messed up by the experience of the accident rather than directly by the physical damage.

But evidence from contemporary patients makes this hypothesis implausible. Damasio describes a patient, Elliot, who had a brain tumor the size of an orange growing in the midline area, which had compressed both the right and left frontal lobes upward. Elliot's life was already falling apart before the tumor was diagnosed. Surgeons spared his life, but sadly not his self. The right prefrontal cortex was destroyed and the midline areas were severely damaged. Bankruptcy, several marriages, and divorces quickly followed Elliot's return to "normal life." He made terrible decisions, showed no foresight, managed time poorly, and let his social relations fall apart. None of this seemed to disturb him.

Surprisingly, Elliot performed well on a wide range of intelligence and personality tests. But on tests for emotions he was a disaster. Things that would have moved him to joy or sorrow in the past were met by flat affect. Eliot in fact did talk about his life before and after the tumor. What was so strange was that he "was able to recount the tragedy of his life with a detachment that was out of step with the magnitude of the events."

On the basis of the evidence provided by Gage, Elliot, and other patients, Damasio hypothesized that such people can reason pretty well up to the point where they need to attach emotional weight to the options before them. They retain knowledge of etiquette and ethics. What they can't do is attach normal affective weights to their choices. They still know, or at least can say on pencil and paper tests, what is good and bad, in good or poor taste, and what shows good or bad sense. But this knowledge doesn't attach in the right ways to their choice making. Reasoning requires the emotions, and the areas in the frontal cortex involved in intention and planning need to communicate with many different brain areas (which contain what Damasio calls "somatic markers" that assign affective color to memories—"these things are good," "these are not good"), including areas all the way down to the brain stem, to arrive at good reasons and appropriate attitudes. Damage to midline areas in the prefrontal areas will severely compromise these abilities. Thus an individual with such damage will no longer be himself.

Damage to brain tissue or biochemical changes due to changes in neurotransmitter levels can change affect, personality, and the self in any number of ways. From the point of view of the naturalist, the defender of the scientific image, when you change, whether it is a lot or a little, there has been a change in your brain. Selves are natural. We are fully embodied animals.

Selfways

There is one last piece of business for this chapter. The position that we are fully embodied animals for whom the brain is the seat of selfhood can be misunderstood to mean that the brain works on its own, as it were, to produce the self. This is a badly mistaken view. We are not born with a self. The self develops, and acquiring it requires living in the world. The world within which our selves emerge includes both natural and social environments. It is primarily the latter that causes us to become the self we are, to see our self the way we do, to describe our self the way we do, to see certain aspects of our self clearly and to occlude from view other aspects. Before we develop much in the way of powers of self-authorship, our families and the culture they are embedded in have most of the say in how we become who we become. Of course, each child starts life with certain temperamental, cognitive, and conative traits that constitute the raw materials his family must deal with as it works to shape him. There comes a time when the individual has developed and internalized certain ways of being an agent himself. At this time, or better, over this period of time he starts to exert and possibly to take over authorial control. I like to think of selves as co-authored, where authorial control shifts from being largely out of one's hands as an infant to being more (possibly largely) in one's control by the time one is an adult.

Cognitive anthropologists are interested in the ways culture forms mind. Each sociocultural environment provides a set of approved models of being a person. These "selfways" incorporate ways of feeling, thinking, and action as well as a set of moral norms each individual is expected to internalize and abide. Research has shown that two important forces (but not the only forces) in constructing selves according to "selfways" are the emotions and morals.

Work in cognitive science has established that, all else being equal, people remember things best that are experienced as emotionally charged at the time of their occurrence. Such memories are more likely than memories with low emotional valence to be seen as causally important in one's life. Furthermore, and this insight comes primarily from work in social science, we know that the narrative models on which people pin their self-conceptions, narratives, and the like are constructed around norms that, at a minimum, involve identifying those aspects of one's self that are most conducive to self-approval and social approval, and that are most relevant in explaining and predicting how I am likely to act in important situations.

Certain cognitive social psychologists in the 1970s posited a mechanism called the "self-serving bias." There exists strong evidence that virtually all people perceive their own performance, physical appearance, and the like as better than average. Except for mildly (but not severely) depressed people, almost everyone is prone to "positive illusions." (That is, mildly depressed people have a more realistic conception of things than optimists or very depressed folk.) Even if a person is told the base rates for divorce, car accidents, and illnesses of various sorts, they predict that these bad things will not happen to them. This work led, within the cognitivist community, to the hypothesis that there was a natural tendency to positively spin one's narrative, which was thought to be tied in turn to a universal need to feel positively about one's self.

More recent work in psychological anthropology has challenged this hypothesis. Steven Heine, Hazel Markus, Darrin Lehman, and Shinobu Kitayama report that Japanese subjects are not prone to the "positive illusions" of the sort North Americans are. They do not systematically bloat their self-assessments in epistemically unjustified ways. Why? The Japanese abide different cultural expectations for being a person. North Americans, the authors write, value "independence, freedom, choice, ability, individual control, individual responsibility, personal expression, success, and happiness," whereas "Japanese style self-esteem requires a comprehensive grasp of another set of core cultural concepts . . . [that] include self-criticism, self-discipline, effort, perseverance, the importance of others, shame and apologies, and balance and emotional restraint." Whereas for North Americans it is important to stand out from the crowd, for Japanese it is not.

Besides demonstrating the importance of culture in determining how the self is constructed, this work shows how a certain sort of self-conception makes individuals who abide it prone to making cognitive mistakes. Does it show that there is no such universal cognitive mechanism as the "self-serving bias"? The answer to this question is less clear. A paper by Heine et al. entitled "Is There a Universal Need for Positive Self-Regard?" can be read as suggesting that the answer is "no." If fact, the authors claim that "the need for positive self-regard, as it is currently conceptualized, is not a universal." North Americans have a strong need for positive self-regard but the Japanese do not. Now this is exactly the sort of profitable disagreement that one can expect to see if social science and cognitive science interact by paying attention to each other. It involves competing claims from cognitive science on one side and cultural anthropology on the other. And it needs to be adjudicated.

But from where I sit—at this philosopher's table—it is not clear that the inference that there is not a universal need for positive self-regard is the right one to take away from the data. North American and Japanese self-conceptions are built around different character traits, to be sure, and one—the North American—leads to more mistakes in self-assessment and prediction on certain tasks. But both of these facts are compatible with the Japanese being as concerned as North Americans are with maintaining positive self-esteem. Their self-esteem simply involves abiding a different set of norms. And these norms, by not supporting a bloated sense of uniqueness, individuality, and immunity to bad luck, do not lead the Japanese to make the same kinds of odd judgments—rightly dubbed "positive illusions"—that North Americans make. Still, there is no evidence that the Japanese are any less concerned than North Americans with personal pride, self-regard, and self-esteem. It is just that these things are achieved by conformity to a less individualistic model.

The concept of self-esteem involves feelings, so if there is a universal cognitive mechanism that pulls for self-esteem it involves essentially the emotions. Almost every contemporary cognitive scientist will admit that we are cognitive-conative animals and that emotions play a role in most if not all of life's tasks. Furthermore, feeling good about oneself involves attachment to a certain normative conception and abiding it. North Americans and Japanese—and everyone else, I am willing to wager—build, abide, and live within conceptions of the self that make them feel worthy, good about themselves.

The culture prizes certain ways of being and living, deems certain ways of conceiving of the self and of being a person as worthy, and judges certain activities and professions as well-suited to living a good and meaningful life. The social setting that sets out the norms for a good life antedates our appearance on earth, and it powerfully constrains who we can or will become, as well as where prospects for meaning and happiness lie. With cultures growing ever more fluid, new or revised norms for living a good human life change during a lifetime. Nonetheless, like all animals, we occupy distinctive niches. The niches suitable for human habitation are significantly more varied than those of most other animals. That said, if a self is to emerge, become a person, and live a good and meaningful life, it will normally be within constraints set out by the social environment in which one lives.

Natural Selves and Natural Endings

In this chapter I have finished the work begun in the previous chapter. I have provided the replacement view for the picture according to which humans are possessed of souls that constitute the self. We are animals possessed of consciousness, including self-consciousness. Self-consciousness is nothing mysterious. It simply involves the capacity to take inner events in consciously in the same way we take in external events by way of our senses. Mother Nature has endowed us with capacities to store, keep track of, and reactivate memories as needed. She has also endowed us with capacities to acquire language—what John Dewey called "the tool of tools." But while language plays a pivotal role in self-representation, we must be careful not to overstate its role in constituting the self. One's sense of identity normally exceeds the limits of what can be captured in words.

Human consciousness is an accumulator. It is accumulative. Besides memories of states of affairs, the mind accumulates a characteristic style of thinking, feeling, acting, and being. When we hold this way of being in view or when we speak of it we are engaging in self-representation. We express ourselves all the time, even when we are not consciously intending to self-reveal. Self-expression is an activity of everchanging, embodied persons, and it both modifies the self and over time results in adjustments to our self-descriptions. Becoming a person, possessing a self, and knowing who one is are necessary conditions for self-control, self-transformation, and self-improvement. Love, friendship, and fulfilling work all depend on who one is, on what one is really like, as do self-esteem and self-respect. Nothing is more important to you and to your friends and loved ones than the person you are.

Someday you will die. Because you are embodied through and through, at that point you will cease to exist. You will not meet death because, as the sage says, "Where death is I am not; where I am death is not, so we never meet." When you die there will no longer be any self that is you. Use your self well while you have it.

SEVEN

Ethics as Human Ecology

> The human good turns out to be the soul's activity that expresses virtue. . . . The end is a sort of living well and doing well in action.
>
> —ARISTOTLE

THERE IS ONE LAST SUBSTANTIAL PIECE of work to do. I've argued that there are ways to understand mind, consciousness, free agency, and selves that leave open the possibility for some sort of conciliation between the manifest and scientific images. The manifest image will have to yield, but will not have to reject its commitments wholesale.

At the start, I said that dis-ease with the scientific image is not simply or only due to the ways it tampers with the concepts of mind, free will, and the self. It comes from the queasy feeling that if the scientific image has its way, the world will have no room left for meaning and morals. We want our lives to have meaning and purpose, and we want the quest to do good and be good to make sense. But if we are just a very smart animal; if our earthly life is all we have; and if everything we believe, think, and value is in large measure the result of contingencies over which we have little control, then what, at the end of the day, is the quest to live a meaningful and good life even about?

We might have the urge to say that certain things matter more than others and that meaning accrues from doing and accomplishing things that matter. And we might have the urge to say it is objectively true that friendship, love, kindness, and compassion are great goods. But what does this even mean? In a theological context, in a framework that deems meaning and morals as set out by God's will, the quest to attain genuine meaning and to find the right moral path makes sense. But if *dasein*, human being, is the way the naturalist paints it, then *dharma*, the project of finding our way, might be without any objective basis or ground. We might find ourselves with urges for meaning and morals, but where in the naturalistic picture do we find guidance? What is the rational basis for our urges for meaning and goodness? Isn't naturalism required to say that human life in fact has no real meaning and that morality, at least as it is commonly conceived, makes no sense? After all, one well-known variant of Darwinism, articulated most famously by Richard Dawkins, says we are simply replication machines for selfish genes.

The naturalist must provide an answer to these questions and quell the associated fears. Here is the answer. Recall I recommended earlier that we think of ethics as a form of ecology. I am using ethics in a broad sense as involving most fundamentally the question: How shall I (we) live? This question intends to locate the grounds for both a meaningful and a moral life, and to provide wisdom about both. Meaning comes from more than morality, but by almost everyone's lights, a life filled with meaningful things lacks something essential if it is not a morally good one. Indeed, it is not implausible to say that being moral is the one necessary condition of a meaningful life. A person might be happy, content, and have succeeded at many of her projects and plans, but if she is immoral we might judge her not to have led a meaningful life. It makes no difference if she judges otherwise. She is wrong. Joseph Goebbels, for instance, was busy and fulfilled for much of his life. His life was meaningful to him perhaps. But it was not meaningful in the sense that requires that goodness objectively penetrates a meaningful life to some significant degree.

The problem for the naturalist is to offer a way of thinking about value, meaning, and worth—moral and nonmoral—that has substance and objectivity. If ethics is conceived broadly, as including meaning and morals, and if it is construed as systematic reflection or inquiry into the conditions required for living a good life, then this can be done. And ecology is the science that studies how living systems relate to each other and to their environment, and so is the relevant analogy. For any natural system, be it a wetland, a coral reef, a population of otters, or the microbial flora of the human large intestine, we can ask what sort of conditions enable the system and its components to flourish.

Ethics, as I conceive it, is systematic inquiry into the conditions (of the world, of individual persons, and of groups of persons) that permit humans to flourish. The formulation might seem to ignore the question of whether humans have responsibilities to nature or to non-human sentient beings. It is intended not to do this. It is my view that if we understand our natures and that of the rest of nature's bounty deeply enough, we will be moved to be morally attentive to the well-being of much more than our fellow humans. We will have moral impulses to care for nature as such. Furthermore, the impulse to care for all of nature will not rest on solely instrumental reasons. It will not spring from enlightened self-interest but from a recognition that the well-being of nature is an intrinsic good. I will not be able to spell out this argument convincingly here. But I say it, first, to reveal what I think is the truth, and, second, to avoid the thought that I take human well-being as the beginning and end-all of ethical inquiry.

Scientism, Again

The task, once again, is to bring the manifest image and scientific image of persons into as harmonious a relation as possible. This can be done only if believers of the manifest image are willing to adjust in certain areas, to yield in their conceptions of mind, free will, and the self, and replace them with ones that fits with the picture advanced by our best science. But there is no need for the manifest image to yield to the *scientistic* variant of the scientific image. *Scientism* is the brash and overreaching doctrine that says that everything worth saying or expressing can be said or expressed in a scientific idiom. When it comes to persons, scientism typically says that the manifest image must be reduced or eliminated. Let me take each point in turn, for one last time.

The claim that science can explain everything we think, say, and do, that it can, in principle, provide a causal account of human being, needs to be distinguished from the claim that everything can be expressed scientifically. Consider art and music. It is patently crazy to say that the work of Michelangelo, da Vinci, Van Gogh, Cezanne, or Picasso, of Mozart, Chopin, or Schönberg, or of Duke Ellington, John Coltrane, Bob Dylan, or Nirvana could be expressed scientifically.

Assuming something like the best case scenario for science, we might want to say that artistic and musical productions can be analyzed in terms of their physical manifestations (painting in terms of chemistry, say, and music in terms of mathematical relationships). Furthermore, some very complex combination of the culture, individual life, and the brain of some artist might allow for something like an explanation sketch of why he produced the works he did. There is nothing remotely odd about these kinds of scientific investigation of art or music, or of the creative process itself. But although such inquiry takes artistic or musical production as something to be explained, it does not take the production itself as expressing something that can be stated scientifically.

The claim that not everything can be expressed scientifically is not a claim that art, music, poetry, literature, and religious experiences cannot in principle be accounted for scientifically, or that these productions involve magical or mysterious powers. Whatever they express, it is something perfectly human, but the appropriate idiom of expression is not a scientific one. The scientific idiom requires words and, often, mathematical formulae. Painting, sculpture, and music require neither. Indeed, they cannot in principle express what they express in words or mathematical formulae. Therefore, whatever they express is not expressible scientifically. Science often tells us useful things about art—about how artists may have been influenced by scientific ideas, for instance, or about the physiology of perception—but it can provide nothing approaching a complete or even satisfying explanation of any artistic work. Not everything worth expressing can or should be expressed scientifically.

There remains the second threat of scientism, that of reduction or elimination. Strictly speaking, reduction and elimination mean different things, and reduction itself comes in various flavors along a continuum. At the near end of the spectrum is what I'll call analytic or functional reduction. Analytic or functional reduction takes some phenomenon, visual perception for example, and tries to explain how it works by way of the interaction of all the subcomponents that produce a visual perception. The phenomenon to be explained might be "perceiving a red rolling ball." Study of the brain shows that color, motion, and shape are processed by different subsystems, and the analytic functionalist explains how each subsystem works to produce the relevant perception. Analytic or functional reduction explains some phenomenon, but it saves the appearances. It takes for granted that we really do see red rolling balls, just as we say we do; it does not explain them away.

A related type of reduction involves cases where one thing, such as water, is shown to be a second thing, H_2O. In a sense, this type of reduction involves showing that what seem to be two things are in fact one and the same. But this is not quite right, because ordinary water is not identical to H_2O. It contains many other chemicals as well. So there really are two things, ordinary water and pure water, and it is only pure water (which is rare) that is one and the same as H_2O. Here one might claim that scientists are only interested in well-behaved natural kinds that obey strict laws. Ordinary water, because it is not pure, is of only marginal interest to the chemist. She is interested in pure water, in H_2O. Ordinary water is important to ordinary people, and to be sure it consists mostly of H_2O. But to the chemist, ordinary water is a heterogeneous kind.

The next step, the worrisome one, is to say that science does not condone or admit discussion of ordinary water in her lawlike explanations. The kind of water that does scientific work, the kind that figures in the laws one learns in Chemistry 101, is pure water, H_2O. Ordinary, impure water names a seemingly homogeneous kind from the point of view of common sense; but it does not name a scientific kind because it is in fact a heterogeneous hodgepodge. If one takes another step, which is not mandatory, and says that only the kinds approved of by science really exist, then it follows that ordinary water doesn't really exist. To be sure, ordinary water exists in the form of puddles or the stuff that comes out of faucets. But once the true nature of water is discovered and dubbed to be H_2O, these things can legitimately be said to only appear to be water, that is, not to be pure water. Reduction thus might be said not to save the appearances. Ordinary water isn't water. It just seems to be water.[1]

[1]Obviously, applied chemists, guardians of the environment, test ordinary water all the time. They are interested in how pure (and thus safe) it is. And thus what they are looking for is what other chemicals besides H_2O appear in a sample and in what quantities.

If one gets in the mood for reductionism of this sort, which we might call eliminative reductionism, then the scientist's job is taken to involve getting at essences, that is, discovering natural kinds. Ordinary kinds, such as what we call "water," "salt," and "gold" serve as her starting point. But once she discovers the essences of these things—H_2O, NaCl, and the substance with atomic number 79—the original phenomena are normally revealed as not being the real things.

Return to the case of perceiving a "red rolling ball." What I called analytic or functional reduction saved the appearance by explaining how it is that we have unified perceptions of red rolling balls. But one could imagine an eliminative reductionist claiming that while there seems to be a unified perception in such cases, there isn't really. Functional decomposition, let's suppose, reveals that there is a lag of some milliseconds between processing in the color, motion, and shape detection areas. Unified perception is an illusion. In this light, the appearances are not saved.

The full-blooded eliminativist thinks that folk categories such as "mind," "consciousness," and "thinking" constitute a heterogeneous hodgepodge, which at best name a host of phenomena that are better described in neural terms. The original concepts of mind, consciousness, and thinking will go the way of terms like "phlogiston," the "ether," and "witches," for which, once we see things truly, we have no need.

I'll just say this much about reduction and eliminativism. They do constitute threats to the manifest image and, in their strongest forms, to my hope for some sort of accommodation of the manifest image and the scientific image. Many philosophers who want to take a stance against reductive and eliminativist maneuvers call themselves

Of course, since ordinary water will be analyzed as admixtures of H_2O as well as whatever other chemicals the sample contains, it seems fairly easy to see how to avoid, even by the purist's lights, the eliminativist conclusion I sketch. Ordinary water exists, it is the wet stuff we call "water." It is just that that stuff is rarely pure water. But it is still, everywhere, some admixture of H_2O plus other natural kinds allowed by chemistry. The line of reasoning I sketch is still helpful in seeing how one might be attracted to the view that kinds as defined by science match up quite imperfectly with the things in the world that we often take them to be identical to. And thus that if water is H_2O, salt is NaCl, and gold is the substance with atomic number 79, ordinary samples of water, salt, and gold are rarely real or pure instances of these things.

"nonreductive physicalists." My own position could be called nonreductive physicalism insofar as I believe that there are genuine levels of complexity in nature, and that the lowest level is not necessarily more real than the others. On the other hand, I strongly approve of analytic or functional reduction, which I see as innocent and principled. We assume that the surface structure of some phenomenon, the way some experience seems, does not reveal why it seems as it does, and we provide an explanation that goes into the deep and hidden structure of mind or of the world to reveal how the experience emerges as it does.

The threat of eliminative reductionism and eliminativism can be easily tamed. Notice, first, that the eliminative reductionist says that only things that are well behaved scientifically are real. The rest are mere appearances. This seems like a weighty metaphysical statement. What is real, what exists, is what is referred to in scientific laws, namely, natural kinds and their instances. All else is appearance. But ultimately this is simply a matter of stipulative terminology. Does anyone really doubt that what comes out of faucets isn't some real thing? Or that it is not rightly dubbed "water"?

Scientific kinds might truly be said to denote essences. They provide terms for phenomena that are well behaved and can thus be wrestled into scientific laws. But it is an unwarranted stretch to restrict "what is real" to that which can be fitted into scientific typologies of only this strict kind. Although the chemist may only be interested in genuine water, that is, H_2O, the ecologist is interested in impure water. Almost certainly there are no interesting laws that govern the behavior of all kinds of impure water, but there are vast numbers of important facts and generalizations about different kinds of impure water.

We can accept that our folk psychology is loose or, what is different, that it names phenomena at a high level of abstraction. Analysis of mind will take such superordinate terms as "consciousness" and mark (as common sense already does) differences among sensations, perceptions, conscious beliefs, conscious desires, conscious memories, moods, emotions, and so on. We can gain further purchase on these phenomena if we explain them in terms of their neural realizers. But once "consciousness" has been explained more deeply, has it been shown to be illusory? I don't see any reason to say so. Suppose the eliminativist takes the next step and says that experience itself is eliminated once we show how each experience can be fully accounted for in terms of the neural state that objectively realizes it. Here we simply need to remind the greedy eliminativist of what, in the third chapter, I called subjective realism. Experiences exist, and they are realized in certain objective states in the nervous systems of certain fully embodied creatures. Experiences are explained as emergent products of those objective states. They cannot be explained away.

So far I've done two things. First, I've explained that the scientistic doctrine that says that everything worth saying can be expressed scientifically is not worth taking seriously. Second, I've shown that a certain ideology that we might call "mad-dog" reductionism or eliminativism is not to be feared. Mad dogs are few, and their bark is worse than their bite.

The scientific image would be an enemy of ethics if scientism was essential to it. But it isn't. But there is still this ground for worry about bringing the scientific and manifest images into some sort of harmonious relation: Science has nothing to say about how it is best to live. Even conceived nonscientistically, the scientific image's silence on this question—it doesn't even seem to deem it worth asking—can be read as implying that nothing much helpful, sensible, or reasonable can be said in answer to it. This is not true.

Ethics Is Empirical

If we take the analogy of ethical inquiry with ecology seriously, then ethics can be conceived as empirical. That is, ethics is inquiry into the conditions that reliably lead to human flourishing. It does not follow, however, that ethics so conceived be viewed as a science, at least as science is commonly understood. A standard and widespread conception of a science is that it should yield precise descriptions of the phenomena to be explained, causal information, and ideally cause-and-effect laws or at least durable and general statistical regularities.

Physics is considered the paragon example of a mature science. Anthropology and history are seen as good at description—especially at what Clifford Geertz calls "thick description," which involves getting into the heads and lives of the people being studied—but weaker at explanation and prediction. To be sure, the anthropologist or historian can explain why some people think and act as they do, or why a certain election, war, or technological change happened or caused other things to happen. But these causal generalizations are singular and local. They are not derived from general laws, nor do they typically provide a basis from which general laws can be inferred. Still, even though anthropology and history are not "hard" sciences, in that they do not wield causal laws, produce causal explanations, or yield firm predictions, no one doubts that they are empirical. I emphasize this point because human ecology will need to make use of anthropological and historical information.

Ecology aims to describe, explain, and predict the behavior of what it studies. But it is also a normative science. It implicitly trades in oughts. Beavers flourish in such and such environments; therefore, if you want to create a good habitat for beavers, you ought to do this and that. Any practical or applied science is like this, even physics: If you want to send a rocket to the moon, then you ought to do this and that. Human physiology, biochemistry, and anatomy figure most prominently in our lives as normative sciences in the form of medicine.

When we ask what science is we often don't consider applied science, and this causes us to see the scientific image as more limited than it in fact is, as consisting pretty much exclusively of inquiry that describes, explains, and predicts things. Once we bring applied science into view we see science in a new light. Some sciences, or parts of sciences, aim to produce certain ends. They seek to make new things, produce new states of affairs, or repair damage to existing states of affairs. Some ends, such as the launching of a weather satellite, are seen to be possible from inside science. Others, like health, antedate science. Even ends that are discovered inside science are normally discussed in wider public spaces before resources are provided to achieve them. Genetics provides us with the means to do many new things. With respect to some of these, human cloning, for example, there is weighty and widespread public discussion of the pros and cons of using the technologies available to produce the end.

We treat medicine's end as unproblematic. Health is a good, so it is uncontroversial that we should seek to produce and renew it. But even here there are controversies. We now possess technologies that allow people to continue to live while helpless or even unconscious.

And we rightly wonder whether just being alive—as opposed to being alive and healthy—is a good thing.

Ethics, in the broad sense, comes in two stripes. First, there is the everyday activity of seeking what some individual or group of individuals conceives, perhaps unconsciously, as their good. Sometimes the challenge of living so as to gain meaning, and to live morally, takes ordinary people up short and prompts them to reflect on what they, individually or collectively, are doing. They pause to ask whether the way they are living is, in fact, a sensible way to get what they are aiming at—meaningfulness and goodness. Second, perhaps as a natural outgrowth of this ordinary reflection, there is systematic reflection of what is good for humans. Systematic reflection is the philosophical discipline of ethics. If what is truly meaningful and good for humans is discoverable, then ethics is the science that systematically discovers and reveals it.

Seeing ethics in this way, as a science, requires us to hold an expansive view of what "science" is. It must—or at least can—include normative inquiry, that is, inquiry into the conditions that lead to some end being attained. In the case of ethics, the end is the physical and psychological flourishing of humans.

I call attention to the possibility of conceiving of ethics as a science for two reasons. First, it allows us to conceive of ethical inquiry as empirical. Observation of humans over history discovers flourishing to be their aim, and living meaningfully and morally to be conditions of so doing. Ethics aims, through exploration of human nature, to discover the ends for humans that lead to meaningful and moral lives. Conceived as a systematic inquiry, it recommends the ends it uncovers or discovers to be good and worthy. It explains or give reasons for thinking certain ends are better than the alternatives,[2] and it provides wisdom about the means to achieve those ends. What makes such inquiry empirical is that it starts from an understanding of human

[2]One might worry that the scientific image has trouble with "reasons." This is an unwarranted worry. First, it is widely believed that reasons are mental states that serve as causes of human action. So if Jane is told her final paper is due at noon, she has reason to turn it in by noon, and this reason figures in the causal explanation of why she does so. Justificatory reasons of the sort that one sees in the talk of engineers and ethicists, for example, involve talk about what states of affairs are desirable and what the best means to

nature as revealed by evolutionary biology, mind science, sociology, anthropology, and history. So the first reason why it is helpful to conceive of ethics as empirical is that it allows us to use actual observation, rather than revealed and traditional wisdom, to determine what outcomes are most reasonably judged good.

Conceiving of ethics as empirical is important for a second reason. It shows us a way around what many have seen as an intractable impasse between the manifest and scientific images. The manifest image trades in oughts. The scientific image trades in is's. You can't derive "ought" from "is," therefore, science has nothing to say about what matters most. It provides no guidance to answering the most important question: How shall we live? But if we acknowledge that the sciences have normative components, then it becomes clear that they are already telling us a great deal about how we should live—without, in the process, sacrificing any of their objectivity.

As I've said, often an end or goal is uncontroversial. It makes sense to send a weather satellite into orbit, and here is how to do so. Orangutans need forests to live in, therefore if it is good for orangutans to survive, we had better preserve forest habitats. If we are asked to provide reasons for some taken-for-granted end, we typically invoke empirical findings. We explain why, for example, the weather's unpredictability is the cause of various kinds of human misery, and argue that some action such as launching a weather satellite will reduce the unpredictability. Lives will be saved if we can better anticipate hurricanes, tornadoes, typhoons, and so on. The goal of saving lives is not deduced from any set of facts, but it makes sense. It is an empirical fact that humans prefer life to death. Living is a necessary condition for everything else, including the pursuit of meaning and morals. Humans selfishly want to avoid being wiped out by terrible weather events, and if we have a modicum of fellow feeling, we do not wish that others be wiped out either. We have two sorts of reasons, one based on self-interest, the other on fellow feeling, to want better

achieve them are. When the reasoning is judged good, the participants to the discussion can load the reasons articulated and produced onto their motivational circuits and attempt to achieve the desired outcome.

weather predictability. So given that a weather satellite will meet a sensible end, we judge that the expense of launching it is worth it.

Nothing mysterious is involved in means–end reasoning, including cases where we seek to identify the ends that are conducive to meaningful and moral human lives. We survey the terrain, we look at persons who are flourishing or have flourished in the past, and we try to identify the personal and environmental conditions that normally lead to flourishing.

I claim that every great ethical text, every great ethical system that has captured the imagination of Westerners (the same holds for Easterners, but I am writing here about the West), involves this sort of reasoning. Aristotle, Kant, and Mill are widely considered the greatest secular ethicists in the West, and I claim that each can be read as trying to locate, describe, and discover the conditions of meaningful and good human lives. Each seeks to describe and defend a way of being that, if embodied in human life, will lead to an excellent life. Kant, most famously, is wary of what ordinary people think is the right way to live. Ordinary people often don't think, or think enough, about how best to live, and the ethicist's job is to provide the wisdom that ordinary people lack. I want to claim that all three great thinkers are engaged in doing ethics as human ecology.

But when I mention Kant, the astute reader will immediately object: Ecology, as you conceive it, is a kind of empirical inquiry, but Kant, in addition to resisting deriving ethics from the lives and "disgusting chit chat" of ordinary folk, explicitly claims to derive morals from a "faculty of pure practical reason." His system is rationalist but not empirical.

Here is my reply. First, ethics conceived as human ecology requires reasoning about the facts, so the deployment of reason is taken for granted. Second, Kant was mistaken in thinking he was deriving the principles of morals from "pure practical reason." Naturalistic philosophy of mind has no room for such a faculty. Naturalistic philosophy is, however, perfectly comfortable with the idea that humans reason, and Kant reasons with a vengeance. What does his reasoning start from and produce? He started from such premises as that we are rational animals who possess the capacities to behave autonomously, to overcome powerful inclinations and temptations.

We aim to be consistent in thought and action. We think of ourselves as possessed of dignity and worthy of respect, and we sensibly desire not to be used by others solely as means to their ends. He concluded that consistency requires us to recognize the dignity of others, to respect them as persons unconditionally, and to permit others to live as they see fit and assist them in living meaningfully and morally insofar as they aim at what is good (or, at least, not at what is bad or unworthy). The reason all this makes sense is because Kant has identified a way of thinking and being that leads to flourishing.[3]

[3]I don't want to insist that every move made by every great moral philosopher is plausibly viewed as falling under the banner of "ethics as human ecology," just as I conceive it. My main point is that more can be so conceived than one might initially think. Kant sometimes says that theories of human flourishing by virtue of being vague and multiple are not, or should not be, what the moral philosopher aims to uncover or discover. This stance, plus his insistence that it is one thing to be happy, another to be good, as well as his view that morality is supreme and overrides all other aims or ends, might well justify reading him as concerned with the imperatives of morality alone. And this would lead to the conclusion that he is not an appropriate example of a moral philosopher, like Aristotle, who is trying to provide a general theory of flourishing. I'm happy to accept this standard reading. But I will still insist that the methods Kant, in fact, uses to isolate the moral components of a good life do fit, or can be reconstructed as fitting, under the general rubric of human ecology.

The rationale behind this claim is that he, like Aristotle, does begin with a certain conception of human nature (we are rational animals) and then aims to discover what moral principles would or should govern the nature of a being so conceived. The aim is to enable us to realize our distinctively moral potential given the kinds of beings we are. Reading Kant in the standard way—as trying to ground morality rather than something broader such as flourishing—has more than a textual advantage, that is, it has more than interpretive credibility of his texts to support it. This further advantage is this: It is characteristic of a certain type of ethical reflection, running from Kant to John Rawls, to divide ethical inquiry into the part that provides a theory of what is right and that which deals with happiness and flourishing (often dubbed "the theory of good"). We start with the assumption that justice or rightness trumps goodness. Thus a happy-making career is not good, all things considered, if it requires me to be unfair to others. The essence of liberalism is often described as involving meeting certain universal duties and then being free to do as one wishes to find a satisfying, meaningful life. Rawls, for example, spends the first two parts of *Theory of Justice* on mandatory moral and political demands, and then offers a rich discussion of human flourishing in Part III. It turns out in my own sketch of what ethics as human ecology will answer to the question of "How shall I (one) live?" that morality, narrowly construed, gets pride of place. But I don't mean to start with the assumption that morality in the narrow sense is the trump suit.

The same is true of the Ten Commandments and the Golden Rule. Both consist of good advice that will lead, under normal circumstances, to individual and communal flourishing. But just as the naturalist will resist Kant's account of deriving the principles of morals from pure practical reason, he will resist the claim that the ethical wisdom contained in the Old Testament and the New Testament reflects God's word. In Kant's case, it is most sensible to say that he articulates wisdom about human life already partly gleaned by other Enlightenment thinkers on the conditions that lead to human flourishing. The same is true of the Bible. The Israelites were not getting on very well coveting each other's property and spouses, and getting embroiled in internecine warfare. The Ten Commandments showed them a better way to live. The same goes for the Golden Rule. Jesus was a noble man who had the charisma to be listened to when he observed that the world would be a much better place if people were less selfish and more loving.

Three Burdens of Ethical Ecology

I have framed the aim of ethics conceived as human ecology as providing wisdom about how to live a meaningful and moral life. One objection to this view will be that it makes ethics a less serious or noble enterprise than when it is conceived as a body of theologically inspired truths, or even when it is conceived as a purely rational enterprise. Perhaps this is true. But ethical wisdom has never been discovered through insight into God's mind or into the deliverances of a faculty of pure reason. Why people yearn for ethical wisdom to be clothed as if it issued from supernatural sources, or from a part of humans that is in touch with divinity, is an interesting question. For myself, having been raised to believe that ethics had such sources, I admit to feeling a certain vertigo in having come to think this cannot be so. Whether individuals raised without the original props can get on without vertigo remains to be seen. It would not surprise me if we needed to conceive of ethics as embodying the highest and most important wisdom there is. Nothing less than life's meaning and our goodness turn on finding our way ethically. Even if we recognize certain stories about the origin of ethical value to be false, it is not a mistake to deem the project of being ethical as in a certain sense aimed at achieving our most sublime potential.

A naturalist might voice a different objection about the concept of ethics as human ecology. She might accept the premise that if there is such a thing as ethical wisdom, it must be gleaned from reflection and observation on human-being, on *dasein,* as revealed over historical time and across many environments. Her objection will be that it seems impossible to say, in an uncontroversial or general way, what human flourishing is. To put it differently, it seems impossible to fix "flourishing" in a way that is not relative to a certain type of social environment or a specific habitat. But if flourishing is relative to specific habitats, then ethics is relative and thus ethical relativism will be true. What's wrong with this? Imagine that no Jews live in this country. Even with no Jews around, the Nazis require as a condition of their flourishing that they experience hatred and disgust toward Jews. They would want to exterminate them if there were any around. It doesn't seem right to have to conclude that hating Jews is necessary for the Nazis to flourish and thus that it is ok for them to hate Jews, even worse that it is good for them to hate Jews insofar as it makes their lives meaningful (even if there are no Jews around to be harmed).

The ethical naturalist who conceives of ethics as human ecology has three different problems. The first is the challenge to say something contentful and general about what human flourishing consists of. The second is to respond to the alleged problem of making any conception of flourishing relative to a particular habitat, which might make it impossible to say anything general about the conditions of flourishing. Finally, and worst of all, an acceptance of relativism may require us to approve of ways of being and living that are unmistakably bad but that are suited to flourishing in particular social environments. Call these the problem of characterizing *flourishing,* the problem of *relativity,* and the problem of *repugnancy.* The burden falls on the ethical ecologist to explain how to deal with them. Here goes.

Flourishing

One might think that an ecologist who works on wetlands or pine forests has a pretty easy job of describing a general picture of what counts as flourishing. In these cases, we simply start with a description of a healthy ecosystem, defined as one that contains reasonable levels of biomass (living matter of all kinds) and biological diversity, that is stable and able to resist invasion from outside species, and that is able to recover quickly from injuries caused by storms, droughts, or fires. The nature of the relevant system, forest, or animal population transparently reveals what it means for it to flourish. The ecologist's job is to isolate and discover what environmental conditions keep the system in question healthy. And health is pretty much all flourishing consists of in these cases.

But in the human case there is the matter of what conduces to living a meaningful and moral life. The Muslim conception of a moral life is different from the Christian conception, and both are different from the Hindu or Buddhist conceptions. Looking back over human history, or thrashing through remote forests with an anthropologist's eye, doesn't help matters. There have been, and are at this moment, many, many different conceptions of what makes for a meaningful and good life.

To make matters worse, there is in every human habitat tension between the conditions of a meaningful life and those of a morally good life. A skeptic about my whole approach might say, "Look, Flanagan, you are very well paid. Many of the projects that you see as being meaningful for you, require you to use your financial resources to follow your passion to write articles and books at nights and on weekends, and not take on night or weekend jobs to make ends meet; to travel for relaxation and to play amateur anthropologist. There are people in great need whom you could benefit greatly if you would turn over half your salary. Yet you won't do this. When your aim to live meaningfully conflicts with your aim to live morally, like most people, you let considerations of personal meaningfulness trump morals."

THE CRITIC IS RIGHT. Meaning and morals do not sit poised from the start to be brought into a smoothly harmonious relation. This is a fact about life. But it doesn't thwart the general aim of ethics as human ecology. It simply shows that for a creature with multiple aims, flourishing may involve trade-offs among mutually unsatisfiable goods.

Owen, the well-off professor, has a problem. My aspirations to live meaningfully and my aspirations to live morally do not quite fit together. I try various ways to work out this problem, but they don't produce the ideal solution, where both aspirations are fully realized. Perhaps the ideal solution is not in the cards, and not just for me now, but for no humans ever.[4] Maybe all we can realistically hope for is to diminish the conflict.

Beyond the problem of trade-offs there is the worry that nothing can be said about what human flourishing consists of generally. I don't think this is true. First, we can say, along with every great moralist, that pleasure is good and pain and suffering are bad. The kinds of pain and suffering that are bad, and that interfere with living meaningfully and morally, are uncontroversial and pretty much universally agreed upon. Pleasure is more complex: Not all pleasures are equally good. Moralists recognize higher and lower pleasures, and the writings of the Stoics, the Epicureans, Plato, Aristotle, the Bible, Mill, Kant, and others provide deep insight into what pleasures are the most worthy ones. Once a pleasure has the imprimatur of being "high," we drop the word *pleasure* and dub it a "worthy" or "noble" thing.

I claim that the things that are deemed morally good and bad, or just good and bad constitute a remarkably stable list over time and, to a significant degree, across cultures. Physical pain and suffering are bad. Friendship, love, and compassion are great goods. Murder, assault, rape, and robbery are very bad. Honesty and courage are virtues. Hypocrisy is a vice, as is cruelty. Insofar as what is morally good is one component of human flourishing, we have a pretty good sense of what it consists of.

What about meaningfulness? What kinds of lives are meaningful and what kinds aren't? The first thing we observe is that meaning is multifarious in a way morality is not. Indeed, some will say there just are no invariants. But this, I think, is false. First, across cultures one finds that being moral, that is, being a good person, is considered a

[4]Saints may be examples of persons who have figured out, indeed who embody, the optimal solution. Moral attentiveness dominates their lives, and other goods, although not totally eschewed, are assigned minimal weight, most especially the acquisition of things that bring them more than minimal material comfort and pleasure.

necessary condition of living a meaningful life.[5] As far as I can tell, it is the only absolutely necessary condition.

When the ancient Greeks—the Stoics, the Epicureans, Plato and Aristotle—described the good life and deemed "happiness" to be one of its essential features, they were not using the word in its modern sense of feeling gleeful or self-satisfied. Happiness for them was akin to contentment, and it comes in its pure form only if a person rightly judges herself to be living a virtuous life. For the Greeks, a person might feel happy but not be happy: She might judge herself to be living a virtuous life, and judge wrongly. Conversely, a person can feel happy but not be so. A person of perhaps impossibly high moral standards might feel unhappy, but be happy, insofar as she is as virtuous as any mortal can be judged to be. But in my experience, such people seem to be happy, to flourish, even as they make sacrifices that most of us find unimaginable.

We might say that subjective happiness is normally a good. But it is only a real good when it is deserved, and it is only deserved when the happy individual is, at a minimum, morally decent.

What else is needed, besides moral goodness, to constitute a meaningful life? There is one other invariant. It is true friendship. Aristotle distinguished among three types of friendship: friendships of utility (one dines each Friday with a business associate); friendships of pleasure (one enjoys the sexual company of another, or associates with an acquaintance because both like to play music together); and true friendship—where you love the other for who he or she is.

True friendship is a great good. Thus we judge a human life that lacks true friendships as lacking a necessary ingredient for meaningfulness. I am inclined, however, to say that it is almost necessary, but not quite necessary. A person who does not aim to have excellent friendships is an odd duck, but the aiming for such friendships may be

[5]Suppose the scientist who discovers the cure for cancer is a consistently and thoroughly immoral creep by everyone's lights. In saying that morality is the one necessary condition of a meaningful life, I am committed to saying *his* life lacks meaning *in the sense* under discussion. He fails to flourish personally. It does not follow that he did not produce something of great good or significance. And thus there is no incoherence in most everyone being very glad he had his time on earth.

what is necessary. The world might not cooperate with allowing one to actually succeed at having them.

Although Aristotle rightly thought that commonality of intelligence, interests, and value makes friendship both most likely and most stable, true friendship might occur with anyone. Certain ascetics remove themselves from the possibility of deep relations with other people for unity with what is holy or divine. It seems to me that such lives display the same impulse for communion with others that lies behind the normal desire to find true friends. Like Aristotle and David Hume, however, I would say that this is an odd way to satisfy the urge. Might such a life be meaningful? I am inclined to say that it might be, since it may produce, in a notional or imaginary way, something akin to what true friendship seeks to achieve in more normal lives.

True friendship might occur between spouses or between parents and their children, especially as the children become older. But again we need to notice that the ideal of true friendship leaves open how, in what way, and with whom, one might find true friendship. This is why we ought rightly resist saying that a meaningful human life requires marriage and family. It doesn't.

My friend Matthieu Richard is a Buddhist monk. Matthieu was born in France and trained as a molecular biologist, and now resides in Kathmandu, Nepal, where he is the French interpreter for His Holiness the 14th Dalai Lama. Matthieu is happy, warm, playful—ever excited, yet at the same time relaxed and calm. He is loving and compassionate to the core. In Dharamsala, he once saw my teenage son, Ben, eating unpeeled fruit—a gastrointestinal no-no—and immediately recommended that Ben go off and drink vodka to kill the germs. Buddhist monks don't normally do vodka, but Matthieu is a down-to-earth monk. And he wears compassion and common sense well together. As a Buddhist monk, Matthieu is celibate and thus has no wife or children. But he has many friends, some of whom are very close, Aristotelian friends. Every Buddhist monk I have met has such friendships.

Although Matthieu's life is very different from mine, we both aim, as best we can, to embody two essential goods in our lives. We both live lives that embody commitment to morality and to true friendship, possibly the only two universal conditions of a good life.

Are there other necessary features of meaningful lives? Maybe. John Rawls invokes what he calls "The Aristotelian Principle" in his discussion of life plans and human flourishing. The Aristotelian Principle says:

> Other things equal, human beings enjoy the exercise of their realized capacities (their innate and trained abilities), and this enjoyment increases the more the capacity is realized, or the greater its complexity.[6]

Rawls calls this "a basic principle of motivation" that "accounts for many of our major desires, and explains why we prefer to do some things and not others by constantly exerting an influence over the flow of our activity."

According to Rawls, humans desire certain things for their own sake—"personal affection and friendship, meaningful work and social cooperation, the pursuit of knowledge, and the fashioning and contemplation of beautiful objects." The desire to engage in complex activities rather than only simple ones comes from the fact that humans relish ingenuity, invention, novelty, and the expression of individuality.

It seems to me that the Aristotelian principle expresses a fundamental fact about human psychology. It tells us something deep about the conditions for human flourishing. Meaningful human lives, we can

---

[6]Assent to the Aristotelean principle will depend on how the clause emphasizing "complexity" is interpreted. Many traditions emphasize "simplicity" as a sensible way to lead a good life. There is no automatic conflict, so long as the Aristotelean principle is read as principled and restrained about the quest for complexity. One might think of the spirit of the bow to complexity as akin to Aristotle's view on true friendship. One can have only a few true friendships, otherwise one is spread too thin. Likewise, only a few of one's talents can be developed complexly. If one dedicates oneself wholeheartedly to developing one's musical, artistic, or philosophical talents, there will not be time or energy to develop all of one's other potential talents to a similar degree. What the claim then comes to is that once one starts to develop a talent, one will have urges to improve it, to take it as far as one can. Furthermore, all advocates of simplicity—Puritans and Quakers and Zen Buddhists—write (truly) as if living the simple life is a complex undertaking, requiring difficult work on one's heart and mind, constant vigilance against the temptations for acquisitiveness, communal encouragement, meditation, prayer, and so on.

now say, involve being moral, having true friends, and having opportunities to express our talents, to find meaningful work, to create and live among beautiful things, and to live cooperatively in social environments where we trust each other. If we have all of these things, then we live meaningfully by any reasonable standard. If we have only some of them, we live less meaningfully, and if we lack all of these things, especially the first two, then our life is meaningless.

This general framework allows that there are many, many different ways to live a good life. There are many different social habitats that can, in principle, allow for human flourishing. Furthermore, different people have different talents and interests. Which ones we seek to develop are a complex outcome of nature and nurture, as well as what our social environment favors. For many persons, realizing their complex talents and interests is not in the cards. Some are prevented by their environment from even discovering what talents and interests they have. Others know what they would do, how they would live, if they had the chance, but they don't.

What Marx called "alienation" is the widespread condition of not being able to discover what one wants or not being remotely positioned to achieve it. Alienation is self-estrangement. An alienated individual feels the tug to uncover her potential, whatever it is, but her social environment prevents her from gaining the education or material support that might lead her to discover who she is and what she might be. Alternatively, she may have discovered her talents and interests, but there may be no social institutions in place to realize her talents or interests. Other times, the opportunities exist, but not for her because she must devote all her energies to surviving or caring for others, possibly in squalor.

Some human lives thus lack meaning. If we think that each human life has intrinsic worth and that each person deserves equal chances to live a good life, then it follows that we should work to make the conditions of living meaningfully universally available.

Aristotle saw this, at which point he turned his attention from ethics to politics. From the perspective of ethics as human ecology, certain socioeconomic conditions need to obtain for humans to flourish. These requirements simply are not in place for all people. Education and meaningful work are not universally available. Beauty is hard to find amid squalor. The lack of basic political freedoms, massively unequal distributions of wealth, and ethnic hatred provide dramatically suboptimal environments even for basic decency and true friendship. These conditions require worldwide social, political, and economic reform.

In any case, I claim to have met the challenge of the skeptic to show that one may make general statements about what human flourishing consists of. I have identified three invariants: first, being moral; second, true friendship; and third, the Aristotelian principle. Assuming a life is moral and that it partakes of the great good of friendship, then its meaning is filled out as an individual is able to express and realize her talents and interests.

Relativity

Morals, as traditionally conceived, put a stop to relativism by claiming that God sets down certain rules that are valid because what God desires is by definition good. Naturalistic ethics, as we have seen, can accept that certain wisdom about how best to live is universal. The naturalist simply rejects the idea that what is right or wrong, good or bad, conducive or not conducive to flourishing, is so because God said so. Insofar as there are universal ethical truths, they are so because they state truths about human well-being that obtain across all human habitats.

To be sure, there are many differences of opinion across and within cultures about various moral matters. But not usually about the big ticket items. To repeat, murder, rape, assault, and theft are considered wrong universally. Honesty and courage are virtues. Dishonesty, hypocrisy, and cowardice are vices.

Some form of sexual modesty is generally considered a virtue. But how this virtue is achieved varies considerably. Female arms and legs should be covered in Buddhist countries. Islamic women should cover their bodies and part of their faces. In places where nudity is okay, one will typically notice two things: First, the weather is warm, and, second, the exposed body parts—female breasts—are not sexualized. Even among hunter-gatherer tribes, it is not that there are no social conventions governing displays of sexuality, it is just that the body itself is not (overly) sexualized.

There are also differences among cultures in how various virtues and values are ranked. Among Chinese—one sees this as far back as Confucius's *Analects*—pious respect for parents and other elders is more important than it is in America and Western Europe. Nonetheless, we do encourage respect for parents and elders. It is just that we don't rank such respect as high as a virtue, or, to put it differently, our conception of due respect for parents and elders takes a different form.

Once again the skeptic may object. Human ecology, he might say, faces the following problem: Human habitats develop over time to have different systems of value, these different value systems set the conditions for living morally and meaningfully within these particular cultures, and thus nothing universal can be said about how best to live. There is no objective answer to the question of whether the Chinese or Americans place the right value on piety toward parents and elders.

We should accept this point, but it shouldn't bother us. Piety, both cultures agree, is good; they simply rate its importance differently. But exactly the same thing happens in nonhuman ecology. Virtually every animal lives in a range of places. Among the habitats that support, say, red-winged blackbirds, there are considerable differences in weather, food supply, and so on. Male red-winged blackbirds are territorial and they like to eat mosquitoes, thus they favor spending spring mating seasons in wetlands where mosquitoes are plentiful. How much waterfront property a male red-winged blackbird will claim depends on the local supply of mosquitoes. Young males usually don't get waterfront property, but they claim larger pieces of less desirable property.

Just as different habitats, even within the same square mile, lead to behavioral differences in red-winged blackbirds, different human habitats lead unsurprisingly to certain differences in lifeways. Explaining why and how they do so is the work of anthropology. Within the context of universal moral principles, different cultures place different weights on different virtues and values—not just moral ones, but on meaningful work, on what is beautiful, on what is fun, and so on. I don't see that this is a problem. Indeed, it seems good that different populations of humans seek to find their good in different ways. It contributes to the sort of beauty we normally appreciate in nature's diversity.

Repugnancy

Occasionally we come across a way of feeling, thinking, or being that is awful but that the people who are committed to it think is fine, good, perhaps even a necessary condition for flourishing. This presents the problem of repugnancy, which may be a special problem for human ecology. Even if the thought of rats living in sewage is disgusting, we have to acknowledge that it really is a terrific way of life for the rats.

But there seems to be something objectively wrong about the lives of the Nazis I imagined earlier, who thrive on hating Jews, even if there are no Jews around for them to harm. When humans engage in racist or sexist practices we judge that something is wrong. This is even true in cases where everyone claims to be happy. We should believe it when we hear that certain Muslim women think that they are treated in the right way, that they like the way things are, even though we judge them to be the objects of grossly unfair, sexist practices. Likewise, an Untouchable Hindu might sincerely say that his lot is fine with him—perhaps because he thinks his past *karma* made him an Untouchable in this current, but ephemeral, life, and his next life will go better.

What resources does human ecology have in such cases? We might notice, first, that the usual ways of adjudicating judgments of awfulness are not only unhelpful but part of the problem. If one of the normal ways of making judgments of value involves consulting what God or the gods think, we will discover that the Muslim and Hindu cases are allegedly divinely endorsed. The Nazi case is not. But what are we to say about all three cases, given that the people who abide these repugnant practices find value and meaning in them? They approve of their behavior. The Nazis might even claim that being Jew-haters makes them happy. Unlike the case of the rats who thrive in sewage, we want to say that these practices are not good for the people who abide them, even if they think they are good, and even if they claim that abiding the practices makes them happy.

Several questions arise, especially once we have granted that people find their good in different ways. Supposing we find some practice wrong, how are we to avoid the charge that we are just projecting onto it norms we happen to accept? From what point of view do we make our judgments?

One way to answer all of these questions at once is to point to the possibility of doing "meta-ecology." Meta-ecology involves trying to isolate the conditions of human flourishing that are not just our own, but that apply across human habitats. We have already engaged in meta-ecology in trying to answer the skeptic's challenge to say something general about the conditions of human flourishing, and in responding to his nervousness about relativism.

Martha Nussbaum and Amartya Sen have done important work of this meta-ecological sort. They call their approach the "capabilities approach," and it is designed specifically to make judgments about how well different cultures are providing the conditions necessary for human flourishing. According to this approach, we ask what an average human is capable of achieving if she is given the chance to develop. The key question is not whether she accepts or approves of her way of being and living, but whether she has chances to develop her human capacities.

The reason Nussbaum and Sen emphasize opportunity over acceptance, approval, or even subjective contentment derives from two observations. The first is that the Aristotelian principle of flourishing captures the set of aims or capacities that, if satisfied, would produce a good life. Second, it is a well-confirmed social psychological observation that people do not always see things clearly or truthfully. Certain social systems inculcate beliefs about the station, duties, and worth of particular persons that lead them to accept that they are not due full access to the sort of development that the Aristotelian principle envisions. Such systems produce malformed persons, who, among other things, may fail to see what they legitimately want or need. It is worth emphasizing that everything said so far can be claimed to be based on empirical observation. The truths cited are truths of philosophical anthropology and psychology. They are truths of what I have been calling ethics as human ecology.

Now let me add this claim: People seek to be treated as ends, with dignity and respect, and not as mere means. This claim, like the previous ones, is one that all rational people will accept. Finally, we need to add this: Each person, purely by virtue of being a person, has the same worth as every other person. From this claim we can, with a little extra work, draw out the conclusion that each person has equal right to achieve what the Aristotelian principle claims each person seeks. Now this last premise might be seen as problematic in that it is not based on any observation but is instead a distinctively normative principle that guides our Western liberal way of thinking but is hardly universal or mandatory. The best answer is that it is based on an observation of what a rational human would acknowledge as the most desirable state of affairs.

Consider John Rawls's famous procedure of asking us to go behind a "veil of ignorance" into the "original position," where we have no information about our own means, intelligence, gender, race, sexual orientation, and so on. Rawls claims that in this position we would choose principles of social justice that express the principle of equal worth. We would require that each person be provided the means to satisfy the Aristotelian principle—to attain, to some reasonable degree, her capabilities. In addition, we would favor social and economic equality and oppose discrimination based on circumstances outside one's control—against those of low intelligence and who are physically infirm, female, Jewish, Catholic, Hindu, and so on.[7] From our position behind the veil, we don't know how this discrimination would affect us.

We can think of this exercise as revealing our fundamental intuitions about justice. These intuitions express, as does the Aristotelian principle itself, a basic feature of human psychology—our sense of justice. Of course, not every basic psychological impulse deserves to be treated as worthy or judged as good. But there is nothing remote-

[7]From behind the veil of ignorance rational persons will not conclude that all persons deserve equal treatment. Those who interfere with the liberty of others, those who disregard the worth, dignity, and rights of others, for example, Nazis, murderers, rapists, sexists, and racists, will be judged to be punishable, or otherwise prevented from doing as they wish.

ly odd about taking an inventory of our basic human impulses and making judgments about which impulses deserve to be enhanced and grown and which ones deserve to be moderated, possibly suppressed, for the sake of flourishing.

Let us return to the Hindu or Muslim women who think they are living well and being treated respectfully, and the Untouchable who thinks he gets the treatment, the life chances, he deserves. From the point of view of meta-ecology, we are in a position to say that they do not see things correctly. They do not see things as they would if their vision was not occluded, if they were not, in certain respects, malformed. Can Muslims and Hindus still go on as Muslims and Hindus once some of their attitudes and practices have been judged as unsuited for human flourishing? Yes. Their situation is not all that different from our own. Slavery has only been universally judged to be wrong for a century and a half in America. And the examination of the degree to which our culture is racist and sexist continues to lead to changes in how we live. Sometimes we discover that certain familiar ways of living depend on racist and sexist beliefs and practices. When this happens various song-and-dance routines take place designed to divert attention from the problems, or rationalizations are produced in an attempt to render the practices "not so bad." Once a practice is challenged in this way, changes are made, often much too slowly. But it is rare that some objectionable practice cannot be modified without taking down a whole form of life. The bus continues to run if black people may sit where they wish, the neighborhood remains desirable after Jews buy houses there, the corporation still turns profits with female executives. If a life-form truly depends in an essential way on repugnant practices, then we have no reason to care if abandoning those practices causes the whole form of life to tumble. It was rotten to the core.

There remain our imaginary Nazis. We deal with them similarly. They are malformed in certain respects: They harbor hatred for Jews. But from behind the veil of ignorance, they would never judge the Jews as worthy of unequal treatment, let alone hatred or extermination if they were to find some in their homeland. Furthermore, even though hatred for out-groups is sadly a common feature of many social environments, it is objectively bad. People who harbor hatred rarely flourish. They may feel self-satisfied, even happy, but these feelings are not reliable measures of a good life.

Buddhists consider anger the most dangerous and unworthy emotion, and hatred is a form of anger. Why is anger so bad? First, it harms both the person who experiences and displays it as well as the person who is the object of it. We in the West don't think of anger as Buddhists do, but we also recognize the dangers to the person who harbors anger. Even when some individual does something that causes us to be rightfully angry, we are aware that it is bad to let our feelings or expressions of anger go too far, to overcome us. With Buddhists, we have observed that anger held too long, or anger that runs too deeply, does no good and poisons the heart and mind. The poison affects more than just our behavior toward whomever we are angry at, it affects our well-being and that of others who hardly deserve our ill will. Anger at one's spouse, if not contained to the person and episode that (we'll assume) warrants it, inevitably leaks into one's being and into the world. Such anger can often be seen in one's foul mood throughout the day, in inattentiveness to the needs of one's colleagues, in disrespect toward those totally undeserving of disrespect.

It is inconceivable that the hatred of our imagined Nazi does not poison his being. How exactly the poisoning reveals itself will be determined by circumstance. But it is an expectable outcome of such hatred that it has other sorry effects.

Even if, unimaginably, the Nazi did not display additional signs of poisoning, he makes a cognitive mistake. He thinks he is justified in hating Jews. But no conceivable set of historical conditions could warrant or justify such hatred. It is one of the four noble truths of Buddhism that we will flourish only if, among other things, we see things truthfully. Aristotle too makes seeing things truthfully a necessary condition of a virtuous life. This requires that we not mistakenly see others as less worthy of respect than ourselves. Now we may, in fact, have psychological tendencies to do exactly this, to be selfish and to favor members of our own group. But once again, conceiving of ethics as human ecology gives us reason to moderate these tendencies. Why? Because observation across cultures reveals that love, kindness, compassion, and inclusion work better to foster flourishing than hatred, meanness, cruelty, and exclusion. Furthermore, clear-headed reflection on race, ethnicity, gender, and economic status reveals these differences to be matters of great contingency. As such, they ought to have no weight when it comes to judging the worth of any person. The warrant for the normative judgment, for the "ought" here, does not emerge from thin air, nor is it a divine deliverance. It comes from clear-headed observation of the conditions that normally lead to flourishing, as well as clear-headed thinking about why we ourselves deserve to flourish.

Someone may ask, "Why should I feel constrained to think consistently? Why, in particular, should I think that others have as much right as me to flourish? Why should I not apply different standards to myself than to others?" There is only this answer. Thinking and behaving consistently generally makes things go better than thinking and behaving inconsistently. Furthermore, living selfishly, with only one's own interest in sight, or those of one's in-group, generally makes life go less well than if one is less selfish.

This is not a proof or a demonstration that it is necessarily good to be loving and compassionate. Selfish creeps, xenophobes, sexists, and racists do sometimes flourish, according to some subjective standard of flourishing. But the fact will remain that they fail to see things truthfully, to think consistently, that is, they fail to see the equal worth of all persons. They do not live worthy lives, as judged from a more objective point of view—from a view that recognizes that we are all equally human, deserving of equal respect, and that we wish to flourish and not to be used as mere means.

The question is really, "Why be moral? Why should I be moral not only in the sense of abiding the conventional do's and don'ts of the Ten Commandments and the Golden Rule, but also why shouldn't I step on other people's toes, if doing so helps me to flourish?" It is worth noticing that the amoralist or immoralist never asks, "Why should I seek pleasure or happiness?" He takes for granted that this is his aim and his right. The amoralist or immoralist, however, is told that to have a personally meaningful and moral life he will need to moderate some of his desires and balance his aim for self-fulfillment against the good of others, and he doesn't see why he should do that.[8]

[8]The amoralist, as I conceive him, does not see any good reason to be moral. He will, however, behave in accordance with conventional morality because he happens to have

The naturalist may make two responses to the person who sincerely wonders "Why be moral?" (The nonnaturalist can resort to the handy and often persuasive argument that you had better be moral because God commands that you do so, and if you aren't you will be punished.) The naturalist must appeal either to prudence or to some sort of "moral sense" backed up by the empirical findings of ethics done as a form of human ecology.

If this conception of ethics can respond to the worries about flourishing, relativism, and repugnancy, it will be useful to sketch a naturalistic genealogy of morals. A clearer sense of the function of ethics in the broad sense may give us leverage in responding to the amoralist or immoralist, and also show us what ethics conceived as human ecology can and cannot do. It cannot do everything we want, if what we want includes giving an absolutely firm foundation to ethics and removing every argument open to the amoralist or immoralist who seeks only her own happiness. But it can do everything we can reasonably expect from the fully naturalistic picture of persons that contemporary science advances.

Human Nature

Daniel Dennett calls Thomas Hobbes "the first sociobiologist," because "he saw that there *had* to be a story to be told about how the state came to be created, and how it brought with it something altogether new on the face of the earth: morality."[9]

According to Hobbes's story (which was not based on empirical observation of any aboriginal humans or their primate relatives) there

no countermoral urges, or because he will be punished for moral misbehavior. The immoralist has countermoral urges and carries through on them, perhaps because he is powerful or sly enough to get away with immorality.

[9]See Dennett, *Darwin's Dangerous Idea*, pp. 453–454. Calling Hobbes a "sociobiologist" might be seen as stretching it. Hobbes is normally read as a Galilean, working to surmise what humans would be like if left to their natural motions (as in the law of inertia). But the largely ignored first twelve chapters of *Leviathan* make it clear that Hobbes sees the need to give a physiological account of sensation, perception, and emotion. Needless to say, because *Leviathan* was written over two centuries before Darwin, the physiology is not inspired by the theory of natural selection.

was a time when pure self-interest reigned. It was every ego for itself. Reason, however, quickly surmised that going with the flow of our natural impulses would impede, if not outright defeat, our individual self-interest. And so morality was invented. Moral philosophy, in Hobbes's view, is prudence's child. In a general theory of prudence, morality is the part that pertains to social cooperation, where such prudence provides the conditions for a decent individual life.

The Hobbesian story makes quick work of the transition from a world of psychological egoists to one of prudential moralists. If humans were to follow their natural impulses and do what they really wanted, each would reach for everything he could get his hands on. This would lead to a war of each against each—the worst possible outcome. Happily we are smart enough to see this in advance, so we compromise: Each agrees to take what she wants only insofar as doing so doesn't seriously compromise the aims of others, who are also playing according to the newly invented rules.

David Hume offers a different picture of human nature. We possess, he wrote, both selfish and benevolent impulses. We want to maximize our personal well-being, but we possess a powerful initial dose of "fellow feeling" as well. Thus our personal well-being consists of more than how we individually fare. Hume was a realist who saw that our benevolent instincts, despite being powerful, extend only so far. Our initial, untutored impulses are to love our loved ones, those who care for us and those we care for. The boundaries of the family, and to a diminishing degree whatever wider community expresses care and concern for us and our loved ones, mark the extent of our original benevolent feelings. Hume thought, however, that these feelings, when activated, are so pleasurable that in a situation of material abundance our initial benevolence would "increase tenfold," and "the cautious, jealous virtue of justice" would not even be thought of.

Hume is aware of a type he refers to as "the sensible knave." The sensible knave behaves well when everyone is watching, but sees no reason to do so when he can act with impunity. Hume says, sounding very much like Aristotle, that we can only hope our moral educational practices produce people with consciences that keep them from knavery.

Both Hobbes and Hume start by trying to locate certain original dispositions in persons who live in normal environments, where at least moderate scarcity exists. Both then try to think through how morality would be apt to develop under such conditions, how moral conformity might be inculcated, and, especially in Hume's case, how our positive moral dispositions might be nurtured and grown.

In Hobbes's case, morality is something that had to be invented. This is because without morality there will be the war of each against each, and no one will gain any earthly goods. In Hume's case, morality requires carefully nourishing certain 'proto-moral' seeds in human nature. This is necessary because in any normal environment, our selfish side will engender conflict, and our benevolent impulses will fail to flourish; they will be blunted in their development and this will leave us less contented than we would otherwise be.

With Charles Darwin we get an even richer picture of how morality might develop. His story differs from Hobbes's in two important respects. First, Darwin is more of a Humean than a Hobbesian in his take on human nature. We are egoists but with fellow feeling. We care about the weal and woe of at least some others. Why we care about them is a complex question. But the fact that we are social animals who mate and reproduce is enough to explain why we might have compassionate dispositions, what Hume called "sympathy," toward our children, mates, and close conspecifics. Second, and this follows from the first point, morality is not "something altogether new on the face of the earth." *Homo sapiens,* presumably like their extinct ancestors as well as such closely related species as chimps and bonobos, possess instincts and emotions that are proto-moral. We are born with the germs, at least, of sympathy, fidelity, and courage. We didn't create the relevant instincts and emotions; natural selection did. To state it in modern terms, a genetic disposition toward fellow feeling for one's kin is more conducive to survival and reproduction than a disposition toward complete selfishness.

Most philosophers prefer the Humean–Darwinian picture of human nature to Hobbes's. One might think this has solely to do with the fact that the former picture is more flattering than the latter, but in fact the Humean–Darwinian picture has science on its side. Nonhuman primates do display a social, convivial side quite naturally. Furthermore, this picture makes scientific sense: Individuals possessed of at least a modicum of fellow feeling will do better at dating, mating, and child-rearing, the key ingredients of reproductive success, than individuals who ignore their conspecifics or use them only for their own ends. The Humean–Darwinian view also ties morality to something more than mere prudence.[10]

Our social instincts and proto-moral emotions are there from the start and thus morality has on its agenda, from the very beginning, concern for the welfare of (some) others. How then do we explain the amoralist or immoralist who seriously asks the question "Why be moral?" The best answer is that he has been malformed or that he lacks proper socialization. Presumably, if the Humean–Darwinian picture is to be preferred over Hobbes's, the amoralist or immoralist does see the point in helping his loved ones—if he has any—to flourish. The fact that he sees no reason to extend his moral impulses beyond himself and them means he hasn't been properly taught that being moral, in some wider sense, really will enable him to flourish more fully. When countermoral impulses arise, as they do for everyone, the powerful temptation to do the wrong thing is explained by the twin facts that something we want is within reach and that we have difficulty keeping in view the remote consequences of our actions—including the damage to our self-respect that will come from doing what we know is wrong.

Hobbes's story has *Homo sapiens* moving within the lifetime of the species from amorality—indeed with powerful impulses toward immorality—to a state of morality that is completely new and unique to humans. Darwin's story, as one might expect, is gradualist, in two respects. Humans, thanks largely to the possession of a cognitive-conative economy that was passed on from ancestors, have proto-moral dispositions from the start. Furthermore, these dispositions are

[10]For some philosophers, morality must be more than a prudential theory because, well, that is what morality is. I don't see that we could complain if the truth was that morality was in fact a subset of a general theory of prudence. The advantage, however, that accrues from thinking of our nature as involving pro-social and not merely prudential attitudes comes not from this sort of conceptual demand originating with a preconceived idea of what "morality" means," but from the fact that imputing dispositions of fellow feeling best explains the way(s) that the members of many related species interact with their conspecifics.

adjustable during one's lifetime. Most mammals, and certainly all primates, organize their social lives with and through feelings: selfish feelings and fellow feelings.

But if this is right, then the Humean–Darwinian story has a serious downside, namely that it ties morality too closely to the emotions. Many of the same philosophers who prefer the Humean–Darwinian picture will say that even if Hobbes was wrong about human nature, at least he saw that morality has to do with reason. Darwin explicitly says that sympathy, experience, reason, instruction, and habit are all involved in the development of our moral sense. Hume, of course, thought the same, as did Aristotle. But what Hume and Darwin share in addition is the view that the emotions are essential to morality even when experience, habit, and reason enter the picture. For Hume, and possibly for Darwin as well, moral reason works with and through the emotions, always and forever.

Because I place significant weight on the importance of the emotions in the genealogy of morals, I need to say a bit about why placing almost any weight on the emotions worries many moral philosophers. Here is a true story that I briefly discussed several chapters ago.

Emotivism

A century after Darwin published *On the Origin of Species*—but as far as we know without any inspiration from Darwin—the philosopher A. J. Ayer gave voice to an influential view of ethics called "emotivism." Emotivism says that ethical discourse, despite often occurring in a declarative, descriptive idiom—"Murder is wrong"—is in fact not descriptive at all. Despite sharing the same grammatical form, ethical statements are not remotely like declarative statements such as "Birds fly" or "Water is H_2O" or "force = mass × acceleration."

The declarative form of ethical discourse disguises feelings, emotions, and pro and con attitudes toward acts, traits of character, types of action, persons, and social groups. According to industrial-strength emotivism, ethical theories purport to ground morals in reason, but they fail to do so because ethics does not have a rational basis. We can try to rationalize our moral preferences, but this rationalization is simply a strategic device designed to put others in the mood to like what we like.

The emotive character of ethical discourse transparently reveals itself, for example, in the Ten Commandments. The Ten Commandments are all in the imperative idiom: "Thou shalt not x." Imperatives do not describe states of affairs. They suggest, recommend, cajole, and order. The emotivist argues that we should take the imperative idiom as basic and view ethical statements in the declarative idiom as disguised imperatives. "Murder is wrong," might be analyzed as a transform of "Thou shalt not kill." Of course, one might ask "Why not?" To which, depending on the version of emotivism in question, the answer is something like "I don't approve of murder," "We don't like murder around here," "Boo for murder!"

The nonemotivist rejects this analysis and says that the basis of the commandment "Thou shalt not kill" is that murder *is* wrong. To which the emotivist asks to be shown the things or properties in the world that make murder wrong, beyond the fact that people don't approve of murder. The relevant properties cannot be found in the world. You can show me dead bodies, but you can't show me "wrongness." Ethical utterances, so the argument goes, *project* "wrongness" onto human actions and eventually—possibly very quickly—we come to think that these moral properties are part of the natural fabric of things.

The upshot is that when I say "Murder is wrong," I am expressing the attitude that "I disapprove of murder" and I am asking or demanding that you do so as well. The idea that ethics might be *fully* analyzed in emotive terms raises a host of worries. One is that, since the time of Plato, reason and emotions are alleged to be at war. Emotions are fickle and untamed, whereas Reason is constant. Emotions reveal our animal nature, whereas Reason reveals the part of us created in God's image—or in Plato's case, the part of us in touch with the unchangeable Forms of "The True" and "The Good." A morally good life involves overcoming emotions. Emotions produce or express temptations, and it is overcoming temptation that morality is largely about. Emotions are hard—possibly impossible—to control. Descartes called the emotions "passions," indicating that they are phenomena that occur without the engagement of our Reason or Will.[11] If, on the other hand, moral statements are not simply expressions of feeling but expressions of facts or states of affairs, then Reason can be our guide. Reason can control and regulate our assessment of facts, guide our thoughts logically, yield the truth, and so on.

Finally, emotivism looks, on its face, way too friendly to relativism. Several famous participants in Plato's *Dialogues* argue that what is good or just depends on what people in various locales "feel" or "think" is just or good. Spartans say "war is good," Athenians say "war is bad" (or at least overrated). There is no fact of the matter about the goodness or badness of war, murder, and conquest. There are only different teams—civilizations, nations, states, communities, tribes—that root for different ways of life. Why? Location, location, location.

Sports rooting behavior is an appropriate analogy. The soccer or baseball team most people root for is determined by such utterly contingent facts as where and when one was brought up, and by whom. When I cheer for the Red Sox and boo the Yankees during a game at Fenway Park, I am just giving voice to these contingencies, as are the Yankee fans. The same nonrational contingencies govern moral allegiance. Hindus, Muslims, Christian fundamentalists, Roman Catholics, Jews, and Buddhists have different practices, which they favor over the alternatives. Why? They were brought up to favor them. End of story.

Opponents of emotivism are quick to point out that rooting for a sport's team is very different from giving voice to one's ethical conviction. When I cheer "Yeah, Red Sox!" I am not trying to cajole the Yankee fans to feel as I do. But when I say "Murder is wrong"—even if I am expressing how I feel about murder—I am trying to get everyone else to agree. Why? The opponent of emotivism might say that it is because murder is *objectively* wrong. The emotivist can't say that, and so she needs a reply.

Hume's reply, a partial one, is that "fellow feeling" is an invariant. Fellow feeling is the emotional spring for other dispositions, such as sympathy, empathy, benevolence, and compassion. Even if morals

[11]Descartes thought we could control how and whether we act on a passion but not the passion itself.

involve emotional expression, they at least involve expression of positive emotions. Furthermore, since these emotions are an invariant part of human nature, they can be expected to yield a universal basic structure of morality. Our basic nature may be enough, for example, to explain why all humans quickly came to think that murder is wrong—at least when it happens to those we care about.

Darwin helps fill out this picture. Despite making emotion central to morality, his view considerably waters down the strongest version of emotivism and thereby yields a more credible theory of the degree to which emotions are involved in morality, how experience, habit, and reason can regulate the emotions, what moral "goodness" and "badness" might truly refer to, and how extreme relativism is, in fact, avoided.

The Basic Emotions

In *The Expression of the Emotions in Man and Animals,* Darwin describes how—possibly even before true language evolved—the emotions could have been centrally implicated in the setting down and coordination of norms and in the construction of morality.[12] Although his writings allow a sensible account of the origin of morality in the emotions, they provide no solace to the hard-core emotivist. Nor do they aid proponents of certain forms of cognitivism and moral realism, who think the emotions need to be sequestered in moral reasoning and who look for "objective" moral properties solely in the world-out-there—or, if they have Platonic leanings, in the world-up-there.

According to Darwin, humans are social animals concerned in the first instance with their own fitness and that of their conspecifics. We are equipped with some rudimentary instincts, or proto-virtues—sympathy, fidelity, and courage—that can be shaped by experience and reason.

[12]My overall aim is to use Darwinian theory to create some leverage to break a certain impasse that afflicts modern moral philosophy. Cognitivists think morals are discovered by reason, whereas noncognitivists—of which emotivism is one variety—think that morals, if we can speak of "discovery" at all, are discovered in our emotions. Rationalists think emotions are the chief enemy of morality, whereas antirationalists think that there is no morality without the emotions. Moral realists think there are "objective" moral properties and thus "objective moral facts," whereas antirealists think this is a silly, baseless idea.

Furthermore, certain emotions are universal and what I call "proto-moral"—meaning that long before *Homo sapiens* articulated such things as moral codes, we used our emotions to regulate social life. Our facial expressions and body movements publicized our desires that certain norms be abided and elicited conformity to these norms. The universality of the expression of certain emotions in certain commonly and repetitively occurring kinds of social situations explains why, when we express our desire for normative conformity, our appeal is universal. A smile, a glare, or a sexy posture is understood the world over.

What is a norm? Nothing mysterious. Roughly, norms express evaluations and make appeals that create, protect, or maintain certain desirable practices. When I display anger, I express a desire that you back off. If you get the message you will do so, and if you are smart you will continue to do so in similar situations. Supposing you do, you now govern your behavior by a norm, consciously or unconsciously. So ethical expression—even as it might be imagined to have occurred before we added language to our expressive arsenal—involves emotional expression, but it is not simply a matter of my hurrahing and booing and getting things off my chest. I am communicating with you *about* our interaction. My reactions reveal that I like or don't like something you are doing or appear disposed to do, and I am attempting to convey information about how I'd prefer things to go.

Like the cognitive neuroscientist Antonio Damasio, I am attracted to the view that thinking always involves the emotions, and conversely that our feelings about human relations almost always involve thought. The early humans I have been asking you to imagine are not doing one and then the other, but both at once. They are observing certain modes of interaction between themselves and their conspecifics and judging the quality of these interactions by assessing what is happening and how it makes themselves and others feel. These feelings involve assessments of whether the interactions succeed at producing contentment, security, and flourishing. We endorse norms governing behavior when we judge that abiding these norms leads to social cooperation and coordination.

In the Introduction to the first edition of *Expression,* Darwin notes that with the exception of Herbert Spencer's work, all previous writings on the expression of emotions start from the assumption that human emotions originated *in* humans, probably at creation. For obvious reasons, Darwin rejects this idea, and thus much of the book is taken up with the examination of expression in nonhuman animals. Homologues abound.[13]

Combining Darwin's insights with the important work of Paul Ekman (who, interestingly, originally set out to prove Darwin wrong) a century later, it is safe to say that certain human emotions are universal. Which ones? Fear, anger, surprise, happiness, sadness, disgust, and contempt, for sure.[14] Darwin also lists sympathy, fidelity, and

[13]Darwin had two data sources for his claim that certain human emotions are universal: (1) He showed photographs of people making different facial expressions to British subjects and noted remarkable consistency in judgments of what emotional state these people were in; and (2) he asked colonialist friends and scientists to respond to a series of questions he posed, all of which suggested his strong suspicion that there were certain universal emotions. Even if one argues that Darwin's conclusions about the universality of certain emotional expressions were tainted by using only British subjects and leading his witnesses in the field, there is now utterly convincing independent confirmation of Darwin's view thanks to the work of Paul Ekman and his colleagues.

There are several reasons for saying that basic emotions exist and evolved via natural selection. First, homologues appear in other animals—canines, as well as in our close ancestors—of the emotions we experience and express (whether or to what extent canines feel emotions as opposed to simply make expressions that will produce normative conformity is an issue about which I remain agnostic). Second, in social mammalian species there are characteristic movements of the facial musculature that are recognized for what they are (for example, for the behavioral dispositions they display) by conspecifics who then seem to respond appropriately to the particular display. Third, for the basic emotions of fear, anger, surprise, happiness, sadness, disgust, and contempt (the search continues for reliable evidence for certain other likely suspects, such as embarrassment, jealousy, puzzlement, defiance, and obstinancy), we have or are well on our way to locating better and better physiological markers that distinguish among the different emotional expressions. Fourth, the emotions, or at least the relevant facial expressions, alleged to be universal are in fact recognized across all human societies for what they are.

[14]It is important to point out that the search for basic emotions revolves around the isolation of distinctive behavioral expressions, especially facial expressions. There may be emotions that are basic but that cannot be detected this way. Finding someone sexually attractive can (I am told) be detected by widening pupils. This may be basic, but it involves almost no movement of the facial musculature. Agreeing or disagreeing with someone could also, I guess, be thought of as falling on the side of the emotions, and Darwin among others suspected that nodding universally meant "Yeah, I agree," whereas shaking the head from side-to-side indicated disagreement. This turns out not to be the case. Furthermore, there are, according to Ekman, culturally specific display rules that make emotions harder to detect from facial expressions in certain cultures.

courage, which despite not coming with characteristic movements of the facial musculature, are assumed to exist solely by virtue of our social nature.

Let us accept therefore that there are universal human emotions. They are part of the original equipment with which we enter the world, just like eyes, ears, noses, and hearts. Unlike eyes and ears, however, which are used in the first instance to pick up information, the basic emotions express how we feel and convey information to others about how we would like them to behave.

Basic Emotions and Reactive Attitudes

But what, you might ask, do the basic emotions have to do with morality? If I glare angrily at you when you reach for my stash of roots, what does this have to do with "right" and "wrong," "duty" and "obligation"? The answer is quite a lot.

Remember, I am trying to bring your behavior under normative governance, even though at the imagined time we lack all the latter moral concepts. The question is when and how and from what these moral concepts arose, and Darwin's theory yields a credible, well-confirmed alternative to the Genesis story of creation. Life as we know it not only could have but did evolve, without the work of an Intelligent Designer.

The basic emotions express that I wish you to behave in certain ways, ways that will lead me to survive and possibly to be reproductively successful. It doesn't matter whether I have fitness as a conscious aim. It is enough that what I am doing, and what we are doing as we try to regulate each others' behavior, has as its aim fitness enhancement. Nor does the account entail emotivism. I express my original moral (or "proto-moral") judgments through my emotions. You learn what I aim to express by detecting what my expression means.

What's that? Two things: I like what you are doing, or I don't. My expression conveys an emotion, but at the same time it expresses a judgment about a way of being or behaving (a norm).

But if ethics begins with fitness enhancement, nothing in the Darwinian picture requires, as social life develops and evolves (according to its own rules), that fitness remain the sole aim of either ethics or human life. In a famous 1962 paper called "Freedom and Resentment," P. F. Strawson proposes an account of what he called the "reactive attitudes." Although Strawson had other fish to fry, his analysis helps reveal how emotional expression, of the sort Darwin and Ekman are interested in, provides just the right sort of crane (remember them?) for morality to be hoisted on.

The reactive attitudes are a set of human responses that include indignation, resentment, gratitude, approbation, guilt, shame, pride, hurt feelings, affection and love, and forgiveness. Strawson claimed that (1) the reactive attitudes are part of the normal and original human conative repertoire; (2) the reactive attitudes *express* normal human reactions to acts, traits, dispositions, or whole persons; and (3) the normal expression of the reactive emotions involves interpersonal relations where benevolence or malevolence is displayed or, at least, where they are at stake. The last point might make it seem as if the reactive attitudes are invariably other-regarding. This is a mistake. We often experience and direct reactive attitudes toward ourselves. Reactive attitudes are thus sometimes, possibly often, self-regarding. A reactive attitude—pride, guilt, shame, feeling one is not getting one's due or a feeling of obligation—can be experienced first-personally by virtue of direct social feedback, for example through the expression of indignation or approval on the part of another, or it can be self-applied even as one simply contemplates some act without carrying it out. According to Darwin, from the very start, "we love praise and dread blame" whether such approbation or disapprobation is delivered by ourselves or others.

So the reactive attitudes are self-regarding as well as other-regarding. They are also experienced and expressed vicariously, that is, on behalf of others. Vicarious expression of the reactive attitudes occurs even where there is no actual impending good or bad will directed toward us. How widely our vicarious expressions of reactive attitudes spread is variable and probably highly dependent on moral education and culture. What we do know is that they can spread very widely over space and time.

Contemporary persons, with knowledge of human history or simply through exposure to television news, feel vicarious indignation toward the malevolent acts of people who are long dead or who engage in atrocious practices halfway around the world, and who in neither case present any reasonable threat to us, our loved ones, or our countrymen.

I don't want to claim that the specific *forms* of contemporary reactive attitudes are "original and natural," only, as Darwin believed, that some form of the reactive attitudes is original and natural. Our kind of animal has dispositions to feel and express these attitudes in all three ways—self-regarding, other-regarding, and vicarious. The specific forms the reactive attitudes take, what triggers them, the degree to which the underlying feelings are moderated and adjusted, are all subject to myriad forces of social learning. Darwin writes: "Ultimately our moral sense or conscience becomes a complex sentiment—originating in the social instincts, largely guided by the approbation of our fellow-men, ruled by reason, self-interest, and in later times by deep religious feelings, and confirmed by instruction and habit."

Strawson himself doesn't explicitly state that the reactive attitudes are a product of our evolution. But it is not unfair to attribute to him some such view. He emphasizes that these attitudes are natural, that they appear in some form across all cultures, and that their ubiquity has something to do with our nature as social creatures.

The Reactive Attitudes and Induction

One reason for thinking that Strawson would approve of seeing the reactive attitudes as products of evolution is that in a telling footnote he compares the reactive attitudes to induction. At this point in his essay, Strawson is concerned with a lurking objection—what if someone comes along and says: "Alright, I get that the reactive attitudes come with our nature, even that these attitudes are universally displayed in negotiating moral life. So what? We need a rational justification for these attitudes." Strawson answers: "Compare the question of the *justification* of induction. The human commitment to inductive belief-formation is original, natural, and non-rational (not *ir*rational), in no way something we choose or could give up."

The idea, I take it, is to suggest that what goes for induction also goes for the reactive attitudes—they are "original, natural, non-rational (not *ir*rational), in no way something we . . . could give up." So the reactive attitudes look like they are a basic feature of our kind of animal. And neither induction nor the reactive attitudes, we can surmise, can be given a justification that doesn't turn on or deploy itself. A Darwinian will of course claim to be able to explain why we are built this way. To which the standard response is that such a causal explanation won't justify what needs justifying. It will just say that we have certain equipment. Now the Darwinian will say that any universal trait is probably a biological adaptation. But the critic will rightly point out that this only tells us that when the trait evolved it led to reproductive success. Saying what makes us fit is important, but normative inquiry in epistemology and ethics is not concerned with biological fitness. Epistemology and ethics are concerned with discovering the norms that lead to justified true belief and to living well and morally, respectively. And these norms involve different aims, or at least broader aims, than simple biological fitness.

All true. Strawson's approach does not require that we deny any of this. His question is simply this: If we accept that we have a certain kind of nature because we evolved to have it (giving us reason to think it originally was an adaptation), are there ways to explain how the original equipment can be modified, moderated, suppressed, or enhanced to fit the aims of epistemology and ethics?

Suppose therefore that we permit Strawson's response that the reactive attitudes, like induction, are part of the basic equipment and thus something ethics and epistemology will need to accept as such. We can explain why we have this equipment—they were adaptations at the time they evolved. This much tells us why Mother Nature endowed us with the relevant equipment. Furthermore, if biological fitness is treated as an acceptable aim, then the explanation gets us in the vicinity of justification. If someone asks us to justify having a heart, we can say that it is better (more conducive to survival) than the alternative. Likewise, the reactive attitudes and inductive belief-forming dispositions enhance the fitness of animals like ourselves. Even if we allow this response, however, pressing questions remain.

First, there is the question of whether Strawson's reactive attitudes are basic in the way Darwin's and Ekman's emotions are. This is a complicated issue. Guilt, shame, indignation, gratitude, and resentment are on Strawson's list but not Ekman's. Guilt and shame are on Darwin's list, as are slyness, jealousy, and embarrassment. One obvious tactic here would be to link Strawson's attitudes as closely as possible with the emotions on the Darwin and Ekman lists. One can get a fair distance with this strategy. So, for example, we can say that indignation is a type of anger directed toward malevolent conspecifics; gratitude is one way we express happiness at being well-treated; sadness is the inverse of happiness. Sadness is an emotional response to losing rather than attaining or retaining some good, say, some thing or relationship we desire. Guilt and perhaps shame might be thought of as anger (or in the case of shame, possibly surprise) directed inward. Regret and remorse might be seen as tied to feelings of real or impending loss, and again to anger directed toward oneself for having brought about such a loss. Rather than detail the entire mapping, I will simply assert that one may draw some fairly close links between Darwin and Ekman's basic emotions and Strawson's reactive attitudes. They have apparently honed in on similar if not identical sets of phenomena.

Strawson, revealing his Kantian—post-Enlightenment—streak, sometimes speaks of the reactive attitudes as being activated when respect for my person is not acknowledged or when my dignity is challenged. I doubt that our early ancestors had our modern concept of dignity but they surely had hierarchies and therefore had a concept of status. We Westerners, children of the Enlightenment, do sometimes experience affronts to our dignity. Strawson can make room for this difference between the reactive attitudes of ancestral and modern persons. Although he is not a social constructionist about reactive attitudes—remember he says they are natural and original—he is clear, in a way no one, to my knowledge, has pointed out, that there are individual and cultural differences in everything from the intensity of the experience or display to the conditions that activate various attitudes. For an orthodox Hindu, getting a haircut the day after one's father dies comes close to sacrilege. For Westerners it is a perfectly reasonable thing to do in preparation for any funeral. Still, in both cultures, there are ways of responding to the death of a close relative that will elicit disapproval. Showing up drunk after visiting a brothel, for example, is not acceptable behavior in America on the day Daddy died (and probably not in India either).

If I am right that the set of reactive attitudes lies in close proximity to certain of Darwin and Ekman's basic emotions, then there is an interesting consequence in the offing. It is this: The basic emotions are to some degree moral, or at least proto-moral. One might object that for an emotion to function as a bona fide moral emotion there must be a moral community in place where members hold in their minds a conception of the demands made by that community. I don't think too much turns on this issue. If the word "moral" causes discomfort, call these emotions "proto-moral," a phrase that asserts only that the individuals involved in social interaction prefer to have certain experiences and not others, that there are goods they desire and bad things they dislike, and that purveyors of good or bad things will be the object of positive or negative reactive attitudes. What sort of things will elicit positive or negative reactive attitudes? The Darwinian answer is that, in the first instance, the eliciting circumstances will be ones relevant to survival and thus to reproductive success. This much will stop extreme relativistic outcomes: murder, trashing my shelter, and stealing my mate or the food I need for sustenance will be bad. They will not be bad solely because I don't like these things. My basic emotions are reliable detectors of modes of interaction that affect reproductive fitness.

In the original evolutionary situation, reactive attitudes convey feelings about interpersonal commerce regardless of whether these emotions may legitimately be dubbed moral emotions. Nor need we worry about whether, deep down, we are all rational egoists concerned only for our own good, or are endowed with a heavy dose of benevolence and fellow feeling. It is enough to imagine us having certain likes and dislikes regarding social behavior and living among compatriots who either inadvertently or on purpose do things to activate those likes and dislikes. Strawson's story can thus work on either Hobbesian or Humean–Darwinian assumptions, although we have seen that the latter story has greater scientific plausibility. For sexually reproducing social species, inclusive genetic fitness is usually enhanced if one cares about more than just oneself.

Within this general evolutionary argument, one may make a case for each of the proto-moral emotions individually as an adaptation. Attitudes such as anger and fear, for example, are easy to imagine having credible links to fitness. Closely related species that display these attitudes commonly do so when they are in physical danger. Attitudes such as happiness and sadness might well subserve various types of social interaction that lead to fitness. If I find sex pleasant I will mate (or try to), and I will convey my pleasure, my happiness, possibly gratitude, to whomever it is I find it pleasant to mate with. If I find playing or being with others pleasant I will be concerned (not necessarily consciously) that they fare well.

Dispositions to express reactive attitudes vicariously might easily be imagined to naturally extend to my offspring or to my mate, as well as to any others whose company I find pleasant, who do me good, or who I detect are in a position to do me good. Likewise, assuming I have come to care about certain others (even if for totally selfish reasons, as we say) I will be disposed to experience anger toward those who harm or threaten them. Primatologists often speak of "moralistic aggression" among primates, where the aggressive display can arise when a chimp is directly threatened or when another chimp he cares about is. Finally, an early warning system where we display our emotions facially before we act on them seems like a good strategy for creatures to benefit from making their feelings and desires known with minimal bodily risk.

Adjustable Adaptations

If we accept that Darwin and Ekman's basic emotions, and Strawson's reactive attitudes, provide insight into our natural and original conative equipment, and that this equipment is the basis from which morality is constructed, several questions remain.

- Are the basic emotions still adaptive?
- How *modifiable* are these emotions?
- If they are modifiable, are there any reasons to work to modify, moderate, or otherwise adjust how, when, and

under what circumstances we experience or express these basic reactive attitudes?

Regarding the first question: Even if we take it for granted that the proto-moral emotions, or some subset of them, can be plausibly defended as adaptations according to the criterion that weights most heavily the causal contribution of a trait to fitness in the original evolutionary situation in which the trait evolved and proliferated, it does not follow that the trait is now adaptive, where, in the first instance, the meaning of adaptive is tied to being fitness producing or enhancing now.

The leaves of eucalyptus trees evolved to be glossy. This serves to preserve moisture, which is a good thing for a tree to do if it fights for survival in an arid climate. If, however, the climate in Australia changes to a tropical one, then the adaptation subserving moisture retention is no longer fitness enhancing, and eucalyptus trees will become oversaturated with retained water and rot away. Eventually, without further adaptation, there would be no more eucalyptus trees and no more koala bears that depend on them.

This example's relevance to the case of the proto-moral emotions is straightforward. Even if these emotions were selected for and maintained in the species because they were once adaptive, this does not establish that they remain adaptive in the environmental niches we now occupy.

This is one reason I simply disagree with Paul Ekman when he says that the idea of destructive emotions is an oxymoron.[15] His guiding idea seems to be that an adaptation cannot be destructive. While an adaptation cannot fail to be fitness enhancing at the time it evolves, environments change, and these changes may produce evolutionary anachronisms. The human social and physical environment has changed radically. The reactive attitudes and the basic emotions may well have evolved as adaptations in close ancestors or in us (it makes a difference which it was, about which more shortly), but whether they continue to function as such depends on how tightly

[15]Paul Ekman said this to me in conversation. I believe he now agrees with me that, for the reasons given, there is no contradiction.

bound they are to certain triggers, and the differences between the environments we now live in and the ones in which these adaptations arose. Any trait, given the right kind of environmental change, can become maladaptive. Thus it is not at all inconsistent that an evolved emotional response can be destructive.

The last two questions concern whether the basic emotions are modifiable and, assuming they are to some extent, whether there are reasons to modify them. These are especially important insofar as one reason Darwinism frightens some moral philosophers is that they think that traits delivered by natural selection are unmodifiable. Their worry is unfounded so long as creatures can learn and so long as the natural traits in question are socially or cognitively penetrable (a potentially important difference that I won't fuss over here).[16] In the footnote where Strawson compares the reactive attitudes to inductive information processing, he gives reason for both pause and hope about the modification of reactive attitudes. To remind you—he writes first that induction (the same goes for the reactive attitudes) is "in no way something we choose or could give up"; but then he immediately adds: "Yet rational criticism and reflection can refine standards and their application, supply 'rules for judging of cause and effect.'"

These comments are harbingers of bad news for anyone who favors "overcoming," in any strong sense, our emotions. But no one—not even Buddhists, I have discovered—actually favors completely overcoming the emotions.[17] Furthermore, Strawson himself concedes that the reactive attitudes are subject to forces of cultural learning.

---

[16]The idea of penetrability is best explained through an example. A reflex such as the way the pupil reacts to light cannot be modified (directly) by my desire (cognitively) for it to behave differently than the light compels it to behave, nor is pupil contraction responsive to efforts by society (requests, education, and so on) to behave differently than it is wired to behave.

[17]Buddhism does favor considerable modification of our original emotional tendencies. Strawson can be read as thinking that the reactive attitudes deserve room to roam in something like their original or natural forms. A view like that of Buddhism requires that the emotions in fact be highly modifiable (unlike reflexes, for example)—that is, that they be cognitively and socially penetrable. Furthermore, it requires the judgment that whatever original fitness-enhancing role the basic emotions had, this role is unnecessary and helpful, possibly harmful, in current cultural settings.

We cannot disentangle emotions from morality, nor should we want to. But we can moderate and adjust our emotions, making them more "apt" to different situations, different social environments, different moral conceptions.

It is plausible that some sort of inductive capacities were selected for because such capacities causally contributed to fitness, to success at hunting, foraging, and mating. Suppose that there was selection for the straight rule of induction: If it is observed that regularity R occurs m/n of the time, infer that it will occur in the future in the same ratio. It is a familiar fact that this rule works fairly well in elementary situations but leads in more complex situations to bad reasoning. As mentioned in an earlier chapter, our intuitive statistical reasoning often breaks down when sample size is too small or when the sample is unrepresentative. The canons of statistics that include norms governing sample size and representativeness are now fairly well known, at least they are taught in methods courses at most universities. As Hans Reichenbach pointed out, these norms themselves were discovered—and it took a long time—by applying the straight rule of induction. The straight rule was selected for and by deploying it in its natural and original form, possibly over many centuries, we came to see or discover ways in which the application of the rule needed to be constrained.

There is, of course, no interesting sense in which the canons of inductive logic, statistics, and probability governing sample size and representativeness were naturally, as opposed to culturally, selected for. But if you aim to accrue firmly grounded knowledge, to do science, or to make accurate predictions in elections, you better apply the relevant canons. In this nonbiological sense the canons are adaptive, as are the abilities to read and write. Applying the norms of good reasoning yields firmer and more accurate knowledge and literacy makes for richer and more pleasant lives than do the alternatives. But neither excellence as a sophisticated reasoner nor literacy is fitness enhancing in the strict biological sense. Indeed, the best predictor in the modern world for low birth rate is the average level of education attained, the two having an inverse relation.

The main point is that a trait, such as being a whiz at applying the sophisticated canons of inductive reasoning or being an avid reader or writer, can be rightly understood as being adaptive in the sense that possessing the relevant trait or ability contributes to knowledge, flourishing, happiness, and the like, without being adaptive in the sense that it contributes to inclusive genetic fitness. Flourishing requires at least that we be fit in the minimal sense of being healthy enough to live a life of reasonable duration. But there came a time when flourishing, living well in the twin senses that involve living meaningfully and living morally, came to be an end in its own right. Indeed, there came a time when it was the most important end, the main condition of a life worth living.

Thus I want to claim the same sort of point that applies to the development of our inductive capacities, as well as whatever other knowledge-producing and -expressing capacities we were originally endowed with for the sake of fitness. These tools that were initially devoted only to fitness enhancement were developed or enhanced to become tools for producing knowledge, literature, for doing science, for expressing our artistic and musical impulses, which in many cases were not, and still are not, solely devoted to the aim of biological fitness. The same applies in the moral case. Some of the reactive emotions on Strawson's list require development, discovery, and canonization over some segment of world historical time. Feelings of pride, dignity, and respect fit this bill. They require, at least as we now understand these concepts, development of a certain conception of a person, of norms governing behavior, do's and don'ts, oughts, institutions governing moral praise and blame, and methods for punishment of those who stray too far from the right path. Relatedly, it might not contribute one iota to my biological fitness if I choose to spread my love and compassion very widely. In fact, I might even choose to decrease my biological fitness if so doing seems worth it. This might happen if how I understand the conditions of flourishing, if how I see my way to living meaningfully and morally, require that I compromise my biological fitness for some greater good. Physicians who go into areas where there are deadly diseases do so knowing they might die. But they feel okay about this. It is worth it. There is no mistake in having come to think and feel this way. There is more to life than biological fitness.

Analogously, induction is applied by the individual or community that has learned to deploy the canons of inductive logic, statistics, and probability theory. But induction is not a ladder we climb and then push away. The same holds for the moral emotions. Our morals express our values, our allegiances, and at no point in moral development do they go away.

One might imagine this objection: Morality involves knowing that one ought to do one's duty even when one doesn't want to. Well, yes. But the right way to think about such cases is by understanding them as cases where we have learned to value, to care about certain things that are not altogether easy to care about or follow through on. The fact remains that if we come to care about doing our duty, we care about it. We are emotionally engaged and invested in doing our duty. And we better be; otherwise we won't do it.

Modifiable Emotions and How Flourishing "Fits" In

Saying that the reactive emotions are natural adaptations is not to say that they evolved in *Homo sapiens.* It is possible that they were delivered pretty much as we are naturally disposed to display them by gene sequences that evolved in earlier groups of hominids or even in some nonhominid ancestor.

Given that our brains and bodies as well as our social structures differ from those of our ancestors, one might think that simply being handed over these ancestors' emotional equipment might not be optimal. Mother Nature cannot evolve structures or behaviors to meet anticipated needs; she can only adapt materials already at hand and in existing environments. Just as tinkering with the panda's wrist bone produced a panda's thumb that was not ideal for picking bamboo leaves, a direct handoff of the reactive attitudes of *Homo erectus,* say, may not have been optimal in either the original evolutionary situation or in the cultured environments humans now live in.

This sort of possibility makes me hesitant to assert wholeheartedly that our original and natural reactive attitudes are still adaptive. Such an assertion would require knowing a whole bunch of things about which I, for one, have only some hunches—about which I am not particularly confident. For example, we will want to know certain things about the standard intensity, if there is such a thing, of the reactive attitudes. How strong or weak are the original settings? We will also want to know what sorts of situations commonly elicit the reactive attitudes. Saying that they are elicited by benevolence and malevolence says little until we know what sorts of things were perceived as benevolent or malevolent. We can make some plausible guesses here. But much information is missing, especially once culture is introduced, because different cultures conceive of benevolence and malevolence somewhat differently.

Still, in current environments we need to be wary of certain of the original and natural reactive attitudes. Expressions of anger in an environment filled with guns has more deadly potential than in a world where the standard expressions can only utilize fists and sticks. Many people once worried, and rightly, that our leaders' apparent willingness to fight a nuclear war that might well have resulted in mass destruction and horrific human loss was caused in part by the facelessness of the enemy. Remember we are attuned to feel anger as well as compassion toward faces, but not toward colored areas on maps. On the other side, mass communication, it is often said, gives face to suffering; we see images of starving children halfway across the globe and are at least sometimes moved to help. In any case, the question of the suitability of reactive attitudes to the modern world is made vastly more complex by the fact that the world is to some significant extent a product of those attitudes.

As for whether the reactive attitudes are modifiable, on the other hand, evidence abounds. Contemporary moral educational practices aim at and sometimes succeed in moderating angry displays, and benevolent dispositions can be developed and enhanced. Different moral communities work in different ways to increase or decrease guilt, inculcate values, and enforce conformity to moral standards.

The aims of morality, or what I've been calling ethics, go beyond setting down norms that enhance or protect fitness. We also aim to live happy, high-quality lives, and to flourish in ways that have virtually nothing (at least directly) to do with fitness. This is where elaborate moral systems come into play. There came a time—every place on earth—when a conception of a person and a conception of a morally good life were discovered, developed, and articulated. Every conception I am familiar with—whether contained in the Torah, in the Old and New Testaments, in Confucius's *Analects,* in the Puranas and the Bhagavad Gita, in the Upanishads, in the Koran, in Buddhist texts, or in the secular moral theories of Aristotle, Mill, and Kant—puts forward all sorts of wisdom and advice about how we ought to structure our cognitive-conative economies, how best to live a life, what virtues are the best expressions of our common humanity, and which feelings and vices we need to be most ready to fight off. Each of these traditions, despite sometimes displaying parochial, xenophobic, sexist, and racist attitudes, identifies problems with living our lives solely according to our biological natures and provides considerable wisdom on how to live a genuinely good life. There is absolutely nothing in Darwinian theory that implies that rational animals such as ourselves can't or don't acquire aims that go beyond fitness. Flourishing is such an aim.

All of these moral traditions constitute evidence that both cultures and individuals can modify ethical responses as the demands of interpersonal life and of human flourishing require.

Conclusion

We are now near the end of our long journey. This chapter was devoted, first, to explaining how to conceive of ethics as an empirical discipline, a form of human ecology.

I explained how conceiving of ethics as human ecology is better than conceiving of ethics either as inquiry into the moral law set down by God or as involving the discovery or uncovering of knowledge we already hold or possess within a faculty of pure reason—knowledge that is already inscribed, in some sense, on our souls. Ethics is best conceived as inquiry into the conditions under which people at different times and in different places flourish.

Humans, I have claimed, are creatures who want to live meaningfully and morally. We have certain natural dispositions that make us want to live this way, and there is nothing remotely objectionable about working to achieve these aims. But they require work, since the demands of living meaningfully and morally are not in natural harmony. Every person who has pursued these aims has experienced the tension between living well and living morally. Conflicts continually arise between the demands of living well personally and of doing what is best for our fellow humans and the other sentient beings that deserve our consideration. This is why ethics, conceived as human ecology, adopts the language of ecology and speaks of balancing and equilibrium. There is no algorithm, however, that can tell us exactly how to balance competing goods.

I have also tried to provide a story about the origins of morality that offers hope that we can continue to become better and better. *Homo sapiens* evolved with biological fitness as its primary aim. As a species, we possess healthy doses of both selfishness and fellow feeling. We carry certain emotional dispositions that equip us to express our feelings. Gradually, over world historical time, we discovered new ends and new means to achieve them. We discovered in ourselves more than simply the need to survive and reproduce, but the need to flourish. That there are now six billion of us attests to the fact that the conditions of biological fitness are now broadly in place. Yet many of our fellows survive in poverty and squalor and have almost no chances to live meaningfully in Rawls's Aristotelian sense, which I will offer again:

> Other things equal, human beings enjoy the exercise of their realized capacities (their innate and trained abilities), and this enjoyment increases the more the capacity is realized, or the greater its complexity.

It is typical, when one's eyes are turned toward those less fortunate, to take one's gaze only as far as the boundaries of one's nation or even one's town or province. This may seem sensible in a situation where it is very hard to see how even to balance all the needs of the citizens of one's own country, let alone others. And thinking of ethics as human ecology might even seem to exacerbate the problem, insofar as it asks us to think of well-being in terms of habitats and in terms of what conduces to flourishing in particular social worlds. But this problem of narrow concern antedates the empirical view of ethics by many centuries. Furthermore, ethics so conceived can help with the problem. Ecology teaches that the health of each ecosystem depends on that of every other ecosystem. It is an empirical mistake to think that a certain ecosystem is in good condition if neighboring ones are not. Even if an ecosystem is in good shape now, its health is unstable so long as neighboring ecosystems are unhealthy. Upon reflection, we easily see that concern for individual ecosystems is shortsighted without concern for the entire world.

In the end, the defender of the scientific image who wants to make peace with the manifest image can make ample room for ethics, for reflective inquiry into living well and living morally. Indeed, because she will not be bound by strictures that prohibit questioning traditional ways of thinking that allegedly carry the imprimatur of God or of pure reason, she has more space for unhindered reflection on the actual conditions under which humans may flourish.

Mind, free will, and the self are not, I have insisted, as the manifest image conceives them. On the other hand, we do have minds, we are persons, and we have capacities to modify and control how we live.

There is, just as Wilfred Sellars hoped, a way to accommodate much of the manifest image within a larger vision that gives space to the scientific image of persons. Some things look different, but much looks as it always did. We are conscious beings on a quest, a quest that achieves its aims when we use our minds to flourish and to be good. These are our most noble aims. They involve striving to become better, individually and collectively, than we are. Insofar as we aim to realize ideals that are possible but not yet real, the quest can be legitimately described as spiritual.

This quest suits the human animal well. It is becoming, worthy, and noble. It is the most we can aim for given the kind of creature we are, and happily it is enough. If you think this is not so, if you want more, if you wish that your life had prospects for transcendent meaning, for more than the personal satisfaction and contentment you can achieve while you are alive, and more than what you will have contributed to the well-being of this world after you die, then you are still in the grip of illusions. Trust me, you can't get more. But what you can get, if you live well, is enough. Don't be greedy. Enough is enough.

BIBLIOGRAPHIC ESSAY

Preface

Wilfred Sellars's 1960 lectures, published as "Philosophy and the Scientific Image of Man" in his *Science, Perception, and Reality,* articulate the overall shape of the problem I discuss in this book—the conflict between the manifest image and the scientific image. I divide Sellars's concept of the manifest image into a human or humanistic image and a world image that pertain respectively to our ordinary everyday conceptions of ourselves, on one side, and of the external world, on the other. My main complaint with Sellars's way of formulating the conflict between the manifest and the scientific image is that he characterizes the "ideal types" of the manifest image, and the perennial philosophy that "refines and endorses it," in completely secular terms. For Sellars, the commonsense picture of "man-in-the world" includes the concepts of consciousness, mind, the self, and free will, but religious concepts and commitments are altogether lacking. One thing about "ideal types" is that one can, within reason, carve them out pretty much as one wishes. But in this particular case, one loses too much of relevance if one carves out the manifest image, and the perennial philosophy "that refines and endorses it," in completely secular terms. The simple reason is that these other concepts—consciousness, mind, the self, and free will—are understood by most ordinary Westerners in ways that reveal theological roots. The dominant concept of mind as a spiritual substance, the self as a soul, and free will as a capacity to circumvent natural law are best understood in terms of the religious concepts they normally support and that are, in turn, supported by them.

A. O. Lovejoy's *The Great Chain of Being* is an excellent book on the still powerful and abiding image that places humans above animals and beneath the angels and God.

Perennial philosophy did become what Sellars sees it as, namely as philosophy *minus* philosophical theology. The secularization of philosophy is a post-Enlightenment phenomenon. At the end of the eighteenth century, a distinction between rational philosophical methods of inquiry and religious and theological ones, between *logos* and *mythos,* was put in place. I do think that there is such a distinction. Furthermore, articulating it allowed philosophy curricula in the new kinds of universities being created in Göttingen and Berlin to become power centers separate from the church. The fact remains that the concepts and categories in whose terms we see ourselves are still there to be analyzed and examined, even if, at the end of the day, they are judged to be themselves, or to be based on, ideas for which there is no rational philosophical warrant. See Don Howard, "The History That We Are: Philosophy as Discipline and the Multiculturalism Debate," for a fine history of the secularization of philosophy in Europe and America.

Chapter 1: Human Being

The guiding idea of this chapter and the next is that we live in a space of images and that images do sometimes sit uncomfortably in something like the same conceptual space. In general, when thoughts, ideas, beliefs, options or feelings are not in harmony, or when all available beliefs or options are unattractive, there is what the psychologist Leon Festinger called "cognitive dissonance." In his 1957 book, *The Theory of Cognitive Dissonance*, Festinger was primarily concerned with the sort of everyday dissonances that come from not wishing to believe that your beloved is both loving and loyal, as she seems, and disloyal, as the evidence suggests (an approach–avoidance conflict), from being accepted to two fine colleges (an approach–approach conflict), or from receiving two lousy job offers (an avoidance–avoidance conflict). But Festinger's basic idea also applies well to the conflicts that arise between *big pictures.* One image I have allegiance to says I am created in God's image, and another to which I also have allegiance dispenses with God talk and says I am the product of natural selection, what Richard Dawkins calls "the blind watchmaker" in his excellent book by that title. This produces dissonance, mental and spiritual cramping.

Martin Heidegger begins his magisterial *Being and Time* (*Sein und Zeit*) this way: "Do we have an answer today to the question of what we really mean by the word 'being' (*seiend*)? Not at all, so it is necessary to pose yet again the question of what 'Being' means." What I take from this is the idea that although we use words like "human being," "man," and "person" as if we knew what we are talking about, we don't really know what the nature of human being is. What I glean from Heidegger's work and definitely agree with are claims such as these: Human nature is not fixed by our biological nature, we are historical beings, beings-in-time; We perhaps uniquely among the creatures on earth seek to find meaning and to live morally; Anxiety normally accompanies human life by virtue of being self-conscious creatures who seek to live meaningfully and morally in a situation in which neither our nature nor the nature of our social world is fixed—and this by virtue of the fact that who we are is a being-in-time.

With respect to Buddhist philosophy I do endorse that philosophy, at least as a good, indeed noble way of thinking, being, and living. The idea of *dharma*, the *Way* to enlightenment, meaning, and righteousness captures a fundamental urge that is also apparent in the West, but that is sometimes denigrated as something about which contemporary academic philosophy has nothing to say.

I didn't need *The New York Times* to convince me that most philosophers don't much like to deal with big questions about life's meaning or about the *Way*. But it may help to convince you that my take on the field is accurate to know that *The New York Times* says so. In an article called "Philosophy in Hiding: I Have Tenure, Therefore I Am" (January, 28, 2001), Peter Edin writes: "Holed up on campus the study of wisdom is struck dumb." Edin bemoans the isolation of philosophy and celebrates the work of Simon Blackburn (see his splendid book *Thinking*) and Alasdair MacIntyre (see *After Virtue*) as two contemporary philosophers who are concerned with wisdom and believe that philosophy has a public role to fill. I'm not sure if Edin is right that there is no philosopher with public visibility in America, or even if it is true in the sense that no philosopher is known in the way some TV or movie stars are, and I'm not sure that the charge of complete isolation in the academy can be made to stick. Several contemporary philosophers, John Rawls and Martha Nussbaum, for example, are very influential in political philosophy, Peter Singer in ethics and bioethics, in particular, and Daniel Dennett in the philosophy of mind and cognitive science.

For any reader interested in the concept of *dharma,* and in particular in its role in defining Eastern philosophy as a wisdom tradition, I recommend E. A. Burtt's classic edition of *The Teachings of the Compassionate Buddha.* Richard King's *Indian Philosophy: An Introduction to Hindu and Buddhist Thought* is a splendid book that explains, among many other things, how the concept of *dharma* fits into the broader framework of Indian epistemology, metaphysics, and ethics.

Chapter 2: The Human Image

Some good books to read to get a better feel for the scientific picture of persons and to get a clear sense of why there is a conflict between the manifest and scientific image are: Charles Darwin's *On the Origin of Species* and *The Descent of Man;* Sigmund Freud's *The Introductory Lectures to Psychoanalysis* and *The Future of an Illusion;* B. F. Skinner's *Science and Human Behavior, Beyond Freedom and Dignity,* and *About Behaviorism;* E. O. Wilson's *Sociobiology,* especially the first and last chapter, as well as *On Human Nature;* Paul Churchland's *The Engine of Reason, the Seat of the Soul;* Daniel Dennett's *Darwin's Dangerous Idea;* Richard Dawkins's *The Selfish Gene;* and Francis Crick's *The Astonishing Hypothesis: The Scientific Search for the Soul.* Not all of these authors are as sympathetic as I am to the prospects of reconciling the two images. For the most part, they don't in fact much discuss the prospects for reconciliation or accommodation.

Two wonderful books on the Copernican revolution are Alexander Koyré's *From a Closed World to an Infinite Universe* and Thomas Kuhn's *The Copernican Revolution.* If you want to start at the beginning read Galileo Galilei's *The Starry Messenger,* and for a compelling introduction to both Ptolemy's theory and Copernicus's theory read his *Dialogue of Two Great World Systems.* This last book reveals Galileo's complete lack of common sense and prudence, and explains why he found himself back under house arrest for the remainder of his life. He has a fellow named Simplicio (the simple-minded one) defend the Ptolemaic view, even though the Pope has warned him explicitly to remain neutral in his presentation.

The best book on the web of belief is called *The Web of Belief* by W. V. O. Quine and Joseph Ullian. A classic on "worldviews" is Stephen Pepper's *World Hypotheses.* Nelson Goodman's *Ways of Worldmaking* is another classic, although Goodman's extreme relativism makes most analytic philosophers very nervous. The idea of a *Weltanschauung* (worldview) made its way into

Western philosophy by way of Germany in the writings of Friedrich Schleiermacher in the nineteenth century and Wilhelm Dilthey at the turn of the last century. Edmund Husserl developed the idea in his concept of the *Lebenswelt* (Life-World), which he used to describe the possibly unconscious background that sets the preconditions for the experiences of historical beings such as ourselves. Heidegger used the term *horizon* to describe what I call the *background*, the implicit set of assumptions we project, again often without full awareness, onto the world. Hermeneutics is the name often given to philosophical work that tries to make explicit the *Weltanschauung*, or *Lebenswelt*, or *horizon* or *background*. Insofar as part of my project involves revealing what the manifest and scientific images consist of, what they assume, I am engaged in hermeneutics.

One of the first books to raise concern about the scientific image in the mid-twentieth century was C. P. Snow's *Two Cultures*. Snow's worry is different from mine, but related to it. He is concerned with the fact that humanists, especially humanities professors who teach great literature, are ignorant of science and suspicious of it. At one level, Snow's essay can be read as a lament about the degree to which the specialization across disciplines leads to problems of cross-disciplinary communication. The problem Snow saw between science and the humanities actually exists within the sciences. I have physicist and mathematician friends who admit to not even understanding what their colleagues do, to not understanding what it is they work on.

Regarding the conflict between science and religion, I recommend Stephen Jay Gould's *Rock of Ages* not because it provides a credible basis for resolving the conflict, but because it reveals how one of the most widely known twentieth-century scientists sees the conflict, which is the way many secular humanists do. There are several really good books on the conflict that takes religious belief seriously: John Polkinghorne's *Belief in God in an Age of Science*; Ian Barbour's *Religion in an Age of Science*; Warren Brown et al.'s (ed.) *What Happened to the Soul?*; and Robert John Russell et al.'s (eds.) *Neuroscience and the Person: Scientific Perspectives on Divine Action*.

Steven Weinberg's essay "A Designer Universe" is an excellent manifesto by a Nobel Prize–winning physicist against theological arguments from design.

The best philosophical essays on ironism are Richard Rorty's in *Contingency, Irony, and Solidarity*. In my experience, literary theorists who take an ironic attitude toward science and especially those who think that science is just as subjective as everything else often mention Andy Pickering's *Constructing Quarks* and Bruno Latour and Steve Woolgar's *Laboratory Life*, about a discovery in endocrinology at the Salk Institute, as evidence for their view. Both of these books are important. But despite situating scientific activity in the real-life circumstances from which knowledge emerges, neither remotely supports the strong view that science is completely a social construction, that is, that it doesn't in some ways and at some points answer to the world. See Ian Hacking's wonderful *The Social Construction of What?* for a look at the weaknesses and superficiality in the actual work that is usually cited by those who claim that science is not in the truth business.

Chapter 3: Mind

Descartes's *Discourse on Method* and his *Meditations*, especially 1, 2, and 6, contain his brilliant defense of mind–body dualism. Gilbert Ryle's *The Concept of Mind* revisits the Cartesian view and claims that the picture of "The Ghost in the Machine" is still the received view, a central component of the manifest image, in the mid-twentieth century. The picture of the ghost in the machine has a deep place in ordinary language, but from Ryle's perspective it is a troublemaker, pure and simple.

When I say that it is a regulative assumption that the mind is the brain, I have left the "is" relation unspecified. How the "is" is to be understood is a thoroughly discussed, but still not completely understood or resolved, problem in the philosophy of mind. Identity theories, functionalist theories, eliminativist theories, and supervenience theories have all been proposed. Paul Churchland's *Matter and Consciousness* is the best overview of the field and the various naturalistic positions, albeit with a strong eliminativist streak. I'm also partial to my *The Science of the Mind* for a historical overview of the philosophical assumptions of contemporary mind science. Many of the seminal articles on identity theory, functionalism, and eliminativism by J. J. C. Smart, U. T. Place, David Armstrong, Jerry Fodor, Hilary Putnam, Daniel Dennett, Donald Davidson, Rob Cummins, and Patricia and Paul Churchland can be found in the following anthologies: Ned Block (ed.), *The Philosophy of Psychology* (two volumes); David Rosenthal (ed.), *The Nature of Mind*; and William G. Lycan (ed.), *Mind and Cognition: A Reader.*

The best historical introduction to the field of cognitive science is Howard Gardner's *The Mind's New Science.* Barbara Von Eckarat's *What Is Cog-*

nitive Science? is an in-depth analysis of the methods and assumptions of computer-inspired cognitive science, that is, pre-cognitive neuroscience from a philosophy of science perspective.

There is an interesting naturalistic position that conceives of mental states very broadly as involving the whole body in interaction with the environment. Excellent work in this "embodied cognition" vein includes that by Francisco Varela, Evan Thompson, and Eleanor Rosch, *The Embodied Mind*; George Lakoff and Mark Johnson, *Philosophy in the Flesh*; and Andy Clark, *Being There: Putting Brain, Body and World Together Again*. Jill Einstein and I defend this sort of view in our "Sexual Identities and Narratives of Self."

Consciousness is currently the sexiest topic in philosophy of mind, as well as in certain quarters of cognitive neuroscience. Although most contemporary philosophers and mind scientists reject Cartesian dualism, the alternative view that human consciousness is a physical phenomenon has been found to be hard to state, let alone prove. The following books and articles will provide a good sense of the hopes, aspirations, problems, and pitfalls associated with trying to rigorously state and defend a naturalistic replacement view for Cartesian dualism: Thomas Nagel, "What Is It Like to be a Bat?"; Charles Marks, *Commissurotomy, Consciousness, and the Unity of Mind*; Joe Levine, "Materialism and Qualia: The Explanatory Gap" and *Purple Haze*; Patricia S. Churchland, *Neurophilosophy*; William G. Lycan, *Consciousness and Experience*; Daniel Dennett, *Consciousness Explained*; John Searle, *Minds, Brains, and Science* and *The Rediscovery of Mind*; Owen Flanagan, *Consciousness Reconsidered* and *Dreaming Souls*; Colin McGinn, "Can We Solve the Mind–Body Problem?", *The Problem of Consciousness*, and *The Mysterious Flame*; Bernard Baars, *A Cognitive Theory of Consciousness* and *In the Theater of Consciousness*; Galen Strawson, *Mental Reality*; David Chalmers, *The Conscious Mind*; Paul M. Churchland, *The Engine of Reason, the Seat of the Soul*; Ned Block, "On a Confusion About the Function of Consciousness"; Frances Crick, *The Astonishing Hypothesis*; Allan Hobson, *Consciousness*; Antonio Damasio, *Descartes' Error* and *The Feeling of What Happens*; Michael Tye, *Ten Problems of Consciousness*; Fred Dretske, *Naturalizing the Mind*; Jaegwon Kim, *Supervenience and Mind*; Charles Siewert, *The Significance of Consciousness*; and John Perry, *Knowledge, Possibility, and Consciousness*.

There are several good anthologies that contain many of the major scientific and philosophical articles on consciousness: Anthony Marcel and Eduardo Bisiach, *Consciousness in Contemporary Society*; Martin Davies and Glyn

Humphrey, *Consciousness*; Thomas Metzinger, *Conscious Experience*; Ned Block, Owen Flanagan, and Güven Güzeldere (eds.), *The Nature of Consciousness*; Stuart Hameroff et al. (eds.), *The Proceedings of the Bi-annual Tucson Discussions and Debates: Towards a Science of Consciousness* (a multivolume series); Masao Ito, Yasushi Miyashita, and Edmund T. Rolls (eds.), *Cognition, Computation, and Consciousness*; Jonathan D. Cohen and Jonathan W. Schooler (eds.), *Scientific Approaches to Consciousness*; and Michael Gazzaniga (ed.), *The Cognitive Neurosciences*.

An excellent anthology on the problem of how to combine first-person phenomenological reports and neuroscience is Jean Petitot et al.'s (eds.) *Naturalizing Phenomenology*, and a nice challenge to scientistic analyses of consciousness, from a scientifically sophisticated philosopher of religion, is B. Alan Wallace's *The Taboo of Subjectivity*.

Chapter 4: Free Will

The useful example of blinking and winking comes from Alice Juarrero. The best place to look for the right view is Aristotle's *Nicomachean Ethics*, book 3. Gilbert Ryle's chapter on "The Will" in *The Concept of Mind* is a provocative challenge to the Cartesian conception, part of Ryle's overall seek-and-destroy mission against Cartesianism.

Among contemporary treatments within the philosophy of mind, Daniel Dennett's *Elbow Room* is an excellent and exciting read. Dennett was a student of Ryle's and Quine's, in reverse order, and the influence shows.

The best statement of agent causation is in Roderick Chisholm's "Human Freedom and the Self" and his *Person and Object*. Robert Kane's *The Significance of Free Will* is the latest noble effort to make sense of agent causation using quantum mechanical ideas. Other interesting defenses of incompatibilism are Peter van Inwagen's *An Essay on Free Will* and Timothy O'Conner's *Persons and Causes*.

The best anthology containing many of the important papers in the twentieth century is Gary Watson's (ed.) *Free Will*.

One way to divide up the impossibly vast literature on the problem is to use the checklist I devised of things we want from the concept of freedom or free agency given that we cannot get free will in the Cartesian sense:

- *Self-control*: Dennett, *Elbow Room*
- *Self-expression and individuality*: Harry Frankfurt, "Freedom of the Will and the Concept of a Person"; Gerald Dworkin, *The Theory and Practice of Autonomy*
- *Reasons-sensitivity*: Dennett, *Elbow Room*; John Martin Fischer and Mark Ravizza, *Responsibility and Control: A Theory of Moral Responsibility*
- *Rational deliberation*: Michael Bratman, *Intention, Plans, and Practical Reason*; Dennett, *Elbow Room*
- *Rational accountability*: Susan Wolf, *Freedom Within Reason*; John Martin Fischer and Mark Ravizza, *Responsibility and Control: A Theory of Moral Responsibility*
- *Moral accountability*: Susan Wolf, *Freedom Within Reason*; John Martin Fischer and Mark Ravizza. *Responsibility and Control: A Theory of Moral Responsibility*; Peter F. Strawson, "Freedom and Resentment"
- *The capacity to do otherwise*: Harry Frankfurt, "Alternate Possibilities and Moral Responsibility"; Dennett, "I Could Not Have Done Otherwise: So What?"; Dennett, *Elbow Room*; A. J. Ayer, "Freedom and Necessity"
- *Unpredictability*: James Gleick, *Chaos*; Alice Juarrero, *Dynamics in Action*; Pierre-Simon Laplace, *A Philosophical Essay on Probabilities*; Barry Loewer, "Freedom from Physics: Quantum Mechanics and Free Will"
- *Political freedom*: Karl Marx, *The Economic and Philosophical Manuscripts of 1844*; John Dewey, *Human Nature and Conduct*

Chapter 5: Permanent Persons

The best overview of the debate about the soul and personal immortality is Anthony Flew's brilliant and comprehensive article "Immortality" in Paul Edwards (ed.), *The Encyclopedia of Philosophy*. Sydney Shoemaker and Richard Swinburne, *Personal Identity: Great Debates in Philosophy*, also nicely lays out the two major contending views.

The doctrine that I am primarily concerned with in this chapter holds that the soul *is* the person, or at least that the soul is the true or essential person. My soul constitutes who and what I really am. The idea that my soul is my essence is tied in the West to the idea that I have prospects for personal immortality. When I die, I don't really die. My essence, my soul, goes on. The classical statement of the view is in Plato's "Phaedo." An alternative view, held, for example, by the Epicurean thinker Lucretius (*De Rerum Natura, III*), is that anything that is born, dies. Plato gets around this idea by arguing that souls preexist the bodies that house them and exist after they die. Platonic souls are eternal. Once the God of Judeo-Christianity appears on the scene, the orthodox view becomes one in which God creates and implants souls at conception (they do not exist prior to God's creative act of implantation), which then go on after the body dies. Thus Judeo-Christian souls are immortal but not eternal.

There are several reasons that I discuss the doctrine of the soul separately from the Cartesian concept of mind. The main one is this: One could believe in Cartesian dualism without believing that my incorporeal mind survives my bodily death. I might in fact be composed of two parts, one physical, the other not. But their fates might be essentially codependent, so that death of either produces dissolution of the other. Aristotle thought this.

Descartes uses the words "mind" and "soul" interchangeably to refer to an incorporeal spiritual substance. The trouble is that, even if one accepts the Cartesian picture of mind, one might have the following sort of thought: My mind is a purely structural feature that takes on a particular character or form as I live my life. Thus at each point in time, I am this person, this embodied soul, with this personality. Conceived in this way, the original undifferentiated soul that breathes life into me takes on its unique character as I live my life and have my experiences. But since my personality changes, this way of thinking seems to make my soul changeable. Some philosophers accept this sort of view and thus accept the doctrine that the part of me that survives my death is my soul so conceived, that is, as including the experiences and personality I acquire. Thus Aquinas writes, "For the soul, even after separation from the body, retains the being which accrues to it when in the body."

In my experience, when ordinary people think of immortality they like to think, at least up to a point, that they survive with their personalities, memories, and so on. Making sense of this prospect is not all that easy however. If my essence is an undifferentiated soul, then perhaps that part, by virtue of being exactly the same over time, and thus immutable and indestructible, survives death and returns to the realm of undifferentiated spirits. But then I don't survive since my personality disappears when my body dies.

This idea has found considerable favor in the East, where *nirvana* is conceived as a state of deliverance from all me-ness. *Nirvana* is a state of me-lessness. Reaching the state of nirvana involves shedding my ego and achieving a state of non-being where all personal memory, desire, and personality are dissolved. But in the individualistic West, the preferred view is one according to which I go on as me, as this person. The problem is to explain how this is possible.

The literature on personal identity, like the literature on mind and free will, is vast. John Perry's anthology *Personal Identity* is the best collection of key classics as well as important contemporary articles. Perry's own "Introduction" sets the debate out beautifully.

Chapter 6: Natural Selves

Two uncontroversially important recent books that develop continuity views of the self are Derek Parfit's *Reasons and Persons* and Kathleen V. Wilkes's *Real People: Personal Identity Without Thought Experiments.*

I have been developing my own view on the nature of the self over the last thirty years. My original inspiration came from reading William James on "The Stream of Thought" in his *Principles of Psychology* and "The Stream of Consciousness" in his *Psychology: The Briefer Course,* in college in the late sixties. Read either chapter; they are pretty much the same. Even then, around the same time as I first read Locke, I had the idea that a naturalist could not think that the self is a permanent, immutable, abiding thing. Readers interested in my views will find them in *The Science of the Mind,* especially in Chapter 2 on William James; *Consciousness Reconsidered,* Chapters 8, 9, and 10, entitled "The Stream of Consciousness," "The Illusion of the Mind's 'I'," and "Consciousness and the Self," respectively; and finally in my *Self-Expressions: Mind, Morals and the Meaning of Life* and *Dreaming Souls: Sleep, Dreams, and the Evolution of the Conscious Mind.*

Daniel Stern's *The Interpersonal World of the Infant* is an excellent book that studies the development of the child's self-concept from a perspective that combines psychoanalytic insights with experimental work in developmental psychology. Charles Taylor's *The Sources of the Self* is a compelling philosophical analysis tracking changes in modernity in the concept of the self. Gary Fireman et al.'s (eds.) *Narrative and Consciousness: Literature, Psychology, and the Brain* is a good recent anthology that brings together work from several disciplines, literature, philosophy, and mind science, on the narrative self. And Donald Spence's *Narrative Truth, Historical Truth* raises really interesting questions about the importance of accuracy in personal narratives. Dennett's "Why Everyone Is a Novelist" presents a fairly strong fictionalist view about the narrative self.

Chapter 7: Ethics as Human Ecology

Aristotle's *Nicomachean Ethics* is the main classical text that inspires my confidence that ethics is best conceived along the lines of human ecology. Like Plato, and every other wise person, Aristotle insists that individual flourishing depends on the existence of a healthy *polis,* a just state. I first explicitly presented my idea of conceiving of ethics as human ecology in "Ethics Naturalized: Ethics as Human Ecology" in Andy Clark, Larry May, and Marilyn Friedman, (eds.) *Mind and Morals,* but the general idea is there in my *Varieties of Moral Personality: Ethics and Psychology Realism.*

Readers interested in the "compatibilities approach," which rests on using an objective standard of flourishing, and not simply the subjective standard of whether people think or feel as if they are living well or that their way of life is good, should read Martha Nussbaum's *Women and Human Development: The Capabilities Approach* and *Sex and Social Justice;* Martha Nussbaum and Amartya Sen's (eds.) *The Quality of Life;* and Amartya Sen's seminal paper "Equality of What?" Nussbaum and Sen use something akin to the "Aristotelian principle of flourishing" discussed in John Rawls's *Theory of Justice* as a way to measure well-being that is not based on standard economic measures such as GNP, nor on people's subjective comfort level or personal assessment of how they are faring.

There are many important articles and books that address the problem of balancing the aims and projects that give life meaning and the aims of morality. These include: Susan Wolf's "Moral Saints"; Robert Adams's "Saints"; Bernard Williams's *Ethics and the Limits of Philosophy* and his essay "Persons, Character and Morality" in *Moral Luck;* Samuel Scheffler's *The Rejection of Consequentialism* and *Human Morality;* Lawrence Becker's *A New Stoicism;* and Shelley Kagan's *The Limits of Morality.* These books will provide the reader with a good sense of the basic structure of the debate, which is typically discussed in terms of the status of "The Overridingness Thesis," which says that if there is a conflict between some personal project or aim—no matter how important or worthy it is—and the demands of morality, morality is trump.

Evolutionary psychology is hot. The field was founded by E. O. Wilson in his classic *Sociobiology: The New Synthesis*, and the field was thus dubbed "sociobiology." Jerome Barlow, Leda Cosmides, and John Tooby (eds.), in *The Adapted Mind*, are responsible for renaming the field "evolutionary psychology." The guiding idea of evolutionary psychology is to see how far modern human psychology can be explained in terms of traits and characteristics that are due to evolution by natural selection, as opposed, say, to those that are due to culture. Much work in this field is quite interesting and respectable. Robert Wright's *The Moral Animal* is a widely and rightly admired work that reveals how much of our own biological natures we must struggle against to be good. However, one has to watch work in the field that tries to derive "ought" from "is," that is, which tries to derive morals from a description of human nature. E. O. Wilson's Pulitzer Prize–winning book *On Human Nature* tries unsuccessfully to ground ethics on Darwinian insights about human nature and to derive certain moral principles from these insights. For an excellent critique of sociobiology and evolutionary psychology insofar as they overreach, see Phillip Kitcher's *Vaulting Ambition*. The *Cambridge Companion to Darwin* contains essays by Kitcher, Alex Rosenberg, and myself, which all, in different ways, raise cautions about evolutionary ethics. My original critique of the overall project of evolutionary ethics is contained in the chapter "Mind, Genes, and Morals: The Case of E. O. Wilson's Sociobiology" in *The Science of the Mind*.

I trust that it is clear that I do not try to use Darwin's theory to derive or ground ethics. My use of Darwin is in the service of providing part of the genealogy of morals. Two excellent works on the question of how our Hobbesian or Humean–Darwinian natures come to accommodate morality are Elliot Sober and David Sloan Wilson's *Unto Others* and Brian Skyrms's *Evolution of the Social Contract*. Also see Richard Dawkins, *The Selfish Gene, The Extended Phenotype*, and *The Blind Watchmaker*.

Daniel Dennett's *Darwin's Dangerous Idea* is a passionate "in your face" manifesto, using Darwinism to challenge prevailing views about human nature. It is the best place to go if one wants to see the conflict between the manifest image and Darwinism painted in stark, possibly the starkest, terms.

For excellent recent work on the emotions, see Paul Griffiths's *What Emotions Really Are* and Paul Ekman's "An Argument for 'Basic' Emotions" and his splendid "Introduction" and "Afterward" to Darwin's classic *The Expression of the Emotions in Man and Animals*. Catherine Lutz's *Unnatural Emotions* is a fascinating work in the social constructionist tradition. The best recent philosophical work in ethics that utilizes Darwinian and Humean insights is Allan Gibbard's *Wise Choices, Apt Feelings*. Simon Blackburn's *Ruling Passions* is another excellent book in the Humean tradition. Gibbard and Blackburn are self-described "non-cognitivists"–"expressivists." I think they get most things right, except what to call themselves, or how to describe their position. Ethical judgments, as I read them, are complex normative judgments that involve inextricably cognitive and conative components.

As far as the classics go, there is Darwin's own *Descent of Man* and *Expression* and P. F. Strawson's "Freedom and Resentment." A. J. Ayer's *Language, Truth, and Logic* contains the original, and still best, defense of emotivism.

The Monk and the Philosopher by Jean-François Revel and Matthieu Ricard is a wonderful dialogue between father, a French political philosopher, and son, a Nepalese Buddhist monk, about human life generally, and includes a rich discussion of the topic of destructive emotions and their role in undermining human flourishing and causing political turmoil. For a discussion of differences between Western and Buddhist views on the emotions see my "Destructive Emotions." See also, of course, the Dalai Lama and Howard Cutler's *The Art of Happiness* and His Holiness's *Ethics for a New Millennium*.

SELECTED BIBLIOGRAPHY

Adams, R. M. 1984. "Saints." *Journal of Philosophy*, 81: 392–401.

Adorno, T. W., et al. 1950. *The Authoritarian Personality*. New York: Harper.

Aristotle. 1985. *Nicomachean Ethics*. Translated by T. H. Irwin. Indianapolis: Hackett Publishing.

Ayer, A. J. 1952. *Language, Truth, and Logic* (2nd ed.). New York: Dover.

______. 1954. "Freedom and Necessity." Reprinted in G. Watson, ed. 1982. *Free Will*. Oxford and New York: Oxford University Press, pp. 15–23.

Baars, B. J. 1988. *A Cognitive Theory of Consciousness*. Cambridge: Cambridge University Press.

______. 1997. *In the Theater of Consciousness: The Workspace of the Mind*. New York and Oxford: Oxford University Press.

Barbour, I. 1990. *Religion in an Age of Science: The Gifford Lectures*, vol.1. San Francisco: Harper.

Barkow, J. H., L. Cosmides, and J. Tooby, eds. 1992. *The Adapted Mind*. New York: Oxford University Press.

Bechtel, W., and G. Graham. 1998. *A Companion to Cognitive Science*. Oxford: Blackwell.

Becker, L. 1998. *A New Stoicism*. Princeton: Princeton University Press.

Berkeley, G. 1996. *Principles of Human Knowledge and Three Dialogues. World's Classics*. New York and Oxford: Oxford University Press.

Blackburn, S. 1998. *Ruling Passions: A Theory of Practical Reasoning*. New York: Clarendon Press.

______. 1999. *Think: A Compelling Introduction to Philosophy*. New York and Oxford: Oxford University Press.

Blakemore, C., and S. Greenfield, eds. 1987. *Mindwaves: Thoughts on Intelligence, Identity, and Consciousness*. Oxford: Blackwell.

Block, N., ed. 1980. *Readings in Philosophy of Psychology* (2 vols.). Cambridge: Harvard University Press.

Block, N. 1997. "On a Confusion About the Function of Consciousness." In N. Block et al., eds. *The Nature of Consciousness: Philosophical Debates*. Cambridge: MIT Press.

Block, N., O. Flanagan, and G. Güzeldere, eds. 1997. *The Nature of Consciousness: Philosophical Debates*. Cambridge: MIT Press.

Boyer, P. 2001. *Religion Explained*. New York: Basic Books.

Brandon, R. N. 1990. *Adaptation and Environment*. Princeton: Princeton University Press.

Bratman, M. 1987. *Intention, Plans, and Practical Reason*. Cambridge: Harvard University Press.

Brown, R. 1965. *Social Psychology*. New York: Free Press/Macmillan.

______. 1986. *Social Psychology* (2nd ed.). New York: Free Press.

Brown, W. S., N. Murphy, and H. Malony, eds. 1998. *Whatever Happened to the Soul?: Scientific and Theological Portraits of Human Nature*. Minneapolis: Fortress Press.

Bruner, J. S. 1983. *In Search of Mind: Essays in Autobiography*. New York: Harper and Row.

Burtt, E. A., ed. 1955. *The Teachings of the Compassionate Buddha: Early Discourses, the Dhammapada, and Later Basic Writings*. New York and Toronto: New American Library.

Calvin, W. 1989. *The Cerebral Symphony: Seashore Reflections on the Structure of Consciousness*. New York: Bantam Books.

Chalmers, D. J. 1995. "Facing Up to the Problem of Consciousness." *Journal of Consciousness Studies*, 2(3): 200–219.

______. 1996. *The Conscious Mind: In Search of a Fundamental Theory*. New York: Oxford University Press.

Chappel, V. C., ed. 1981. *The Philosophy of the Mind*. First printed by Prentice Hall in 1962. New York: Dover Publications.

Chisholm, R. 1964. "Human Freedom and the Self." Reprinted in G. Watson, ed. 1982. *Free Will*. Oxford and New York: Oxford University Press, pp. 24–35.

______. 1976. *Person and Object*. LaSalle, Ill.: Open Court.

Churchland, P. M. 1984. *Matter and Consciousness: A Contemporary Introduction to the Philosophy of Mind*. Cambridge: MIT Press.

______. 1989. *A Neurocomputational Perspective: The Nature of Mind and the Structure of Science*. Cambridge: MIT Press/Bradford Books.

______. 1995. *The Engine of Reason, the Seat of the Soul: A Philosophical Journey into the Brain*. Cambridge: MIT Press.

Churchland, P. S. 1986. *Neurophilosophy: Toward a Unified Science of the Mind/Brain*. Cambridge: MIT Press.

Clark, A. 1997. *Being There*. Cambridge: MIT Press.

Cohen, J., and J. W. Schooler, eds. 1997. *Scientific Approaches to Consciousness*. Mahwah, N.J.: Lawrence Erlbaum Associates.

Crick, F. 1994. *The Astonishing Hypothesis: The Scientific Search for the Soul*. New York: Charles Scribners Sons.

Dalai Lama, 1999. *Ethics for the New Millennium*. New York: Riverhead Books.

Dalai Lama and H. Cutler, 1998. *The Art of Happiness: A Handbook for Living*. New York: Riverhead Books.

Damasio, A. 1994. *Descartes' Error: Emotion, Reason, and the Human Brain*. New York: Avon Books.

______. 1999. *The Feeling of What Happens: Body and Emotion in the Making of Consciousness*. New York: Harcourt Brace & Co.

Darley, J. M., and C. D. Batson. 1973. "From Jerusalem to Jericho: A Study of Situational and Dispositional Variables in Helping Behavior." *Journal of Personality and Social Psychology*, 27: 100–108.

Darwin, C. 1859. *On the Origin of Species*. The Modern Library. New York: Random House.

______. 1871. *The Descent of Man*. The Modern Library. New York: Random House.

______. 1872. *The Expression of the Emotions in Man and Animals* (3rd ed., 1998). New York and Oxford: Oxford University Press.

Davies, M., and G. Humphrey. 1993. *Consciousness: Psychological and Philosophical Essays*. Oxford and Cambridge: Blackwell.

Dawkins, R. 1976. *The Selfish Gene*. New York: Oxford University Press.

______. 1982. *The Extended Phenotype*. San Francisco: Freeman.

______. 1986. *The Blind Watchmaker*. New York: W.W. Norton.

Deacon, T. 1997. *The Symbolic Species: The Co-evolution of Language and the Brain*. New York: W.W. Norton.

Dennett, D. C. 1978. *Brainstorms: Philosophical Essays on Mind & Psychology*. Montgomery, Vt.: Bradford Books.

______. 1984. *Elbow Room: The Varieties of Free Will Worth Wanting*. Cambridge and London: MIT Press.

______. 1984. "I Could Not Have Done Otherwise: So What?" *Journal of Philosophy*, 81(10): 553–565.

______. 1987. *The International Stance*. Cambridge: MIT Press/Bradford Books.

______. 1988. "Why Everyone Is a Novelist." *Times Literary Supplement*, 4, No. 459.

______. 1991. *Consciousness Explained*. Toronto: Little, Brown & Co.

______. 1995. *Darwin's Dangerous Idea: Evolution and the Meanings of Life*. New York: Simon & Schuster.

Descartes, R. 1993. *Discourse on Method and Meditations on First Philosophy*. Translated by D. Cress. Indianapolis: Hackett Publishing.

De Sousa, R. 1987. *The Rationality of Emotion*. Cambridge: MIT Press/Bradford Books.

Dewey, J. 1922. *Human Nature and Conduct*. New York: Henry Holt.

Double, R. 1991. *The Non-Reality of Free Will*. New York: Oxford University Press.

Dretske, F. 1988. *Explaining Behavior*. Cambridge: MIT Press.

______. 1993. "Conscious Experience." *Mind*, 102(406): 263–283.

______. 1995. *Naturalizing the Mind*. Cambridge and London: MIT Press.

Dworkin, G. 1988. *The Theory and Practice of Autonomy*. Cambridge: Cambridge University Press.

Eccles, J. 1991. *Evolution of the Brain: Creation of the Self*. First printed in 1989. London: Routledge.

Eckardt, B. V. 1993. *What Is Cognitive Science?* Cambridge and London: MIT Press.

Edelman, G., and G. Tononi. 2000. *A Universe of Consciousness: How Matter Becomes Imagination*. New York: Basic Books.

Einstein, G., and O. Flanagan. 2002. "Sexual Identities and Narratives of Self." In G. D. Fireman et al. *Narrative and Consciousness: Literature, Psychology, and the Brain*. Oxford: Oxford University Press.

Ekman, P. 1972. *Emotions in the Human Face*. New York: Pergamon.

______. 1992. "Are There Basic Emotions?" *Psychological Review*, 99(3): 550–553.

______. 1998. "Introduction, Afterword, and Commentaries" to C. Darwin. *The Expression of the Emotions in Man and Animals* (3rd ed.). New York and Oxford: Oxford University Press.

Ekman, P., R. W. Levinson, and W. V. Freisen.1985. "Autonomic Nervous System Activity Distinguished Among Emotions." *Science*, 22: 1208–1210.

Erikson, E. H. 1968. *Identity: Youth and Crisis*. New York: W.W. Norton.

Feigl, H. 1967. *The Mental and the Physical*. Cambridge: MIT Press.

Festinger, L. 1957. *A Theory of Cognitive Dissonance*. Evanston, Ill.: Row, Peterson.

Fireman, G. D., T. E. McVay, and O. Flanagan. 2002. *Narrative and Consciousness: Literature, Psychology, and the Brain*. Oxford: Oxford University Press.

Fischer, J. M., and M. Ravizza. 1998. *Responsibility and Control: A Theory of Moral Responsibility*. Cambridge: Cambridge University Press.

Flanagan, O. 1991. *The Science of the Mind*. Cambridge: MIT Press.

______. 1991. *Varieties of Moral Personality: Ethics and Psychological Realism*. Cambridge and London: Harvard University Press.

______. 1992. *Consciousness Reconsidered*. Cambridge: MIT Press.

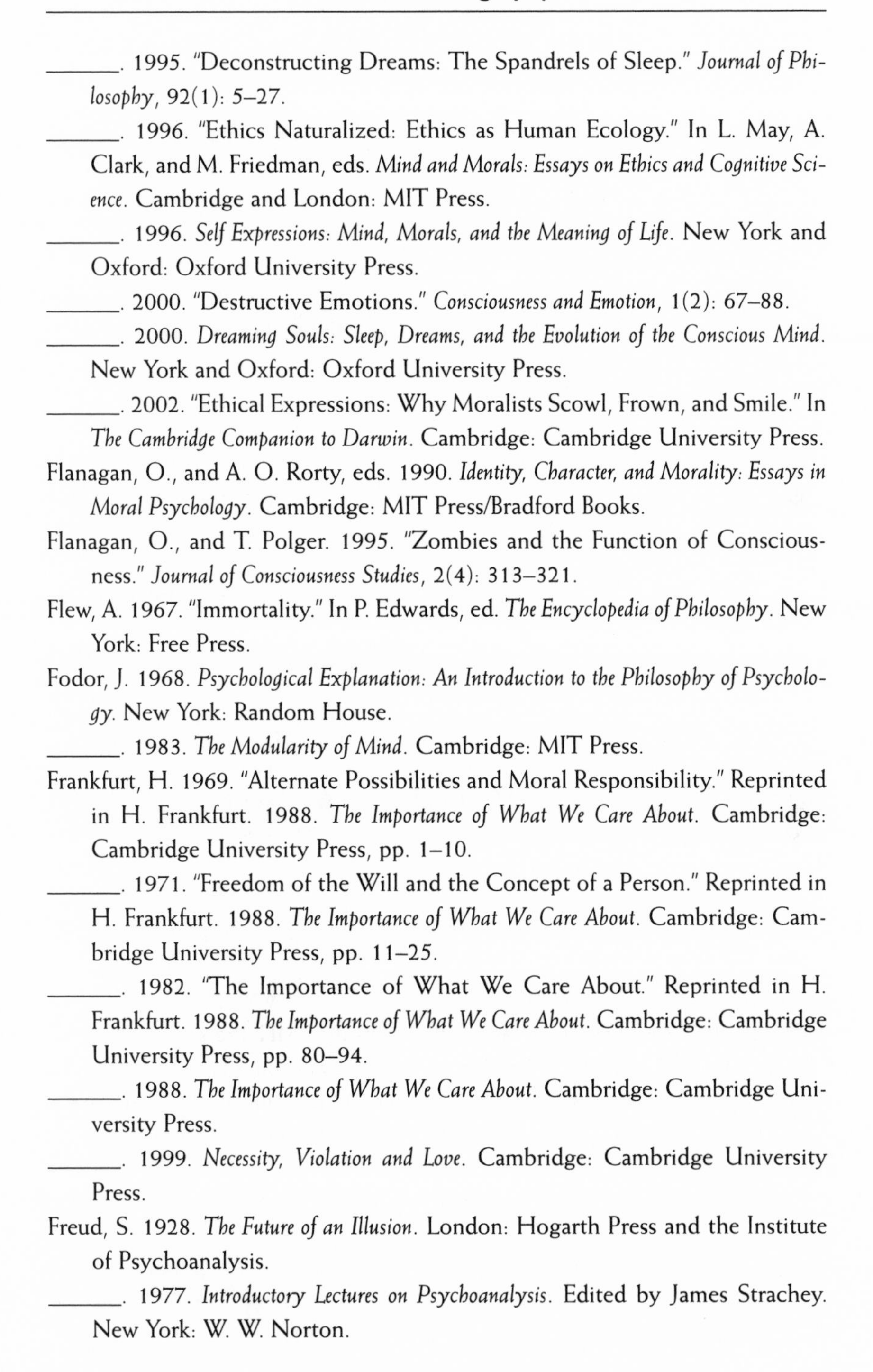

______. 1995. "Deconstructing Dreams: The Spandrels of Sleep." *Journal of Philosophy*, 92(1): 5–27.

______. 1996. "Ethics Naturalized: Ethics as Human Ecology." In L. May, A. Clark, and M. Friedman, eds. *Mind and Morals: Essays on Ethics and Cognitive Science*. Cambridge and London: MIT Press.

______. 1996. *Self Expressions: Mind, Morals, and the Meaning of Life*. New York and Oxford: Oxford University Press.

______. 2000. "Destructive Emotions." *Consciousness and Emotion*, 1(2): 67–88.

______. 2000. *Dreaming Souls: Sleep, Dreams, and the Evolution of the Conscious Mind*. New York and Oxford: Oxford University Press.

______. 2002. "Ethical Expressions: Why Moralists Scowl, Frown, and Smile." In *The Cambridge Companion to Darwin*. Cambridge: Cambridge University Press.

Flanagan, O., and A. O. Rorty, eds. 1990. *Identity, Character, and Morality: Essays in Moral Psychology*. Cambridge: MIT Press/Bradford Books.

Flanagan, O., and T. Polger. 1995. "Zombies and the Function of Consciousness." *Journal of Consciousness Studies*, 2(4): 313–321.

Flew, A. 1967. "Immortality." In P. Edwards, ed. *The Encyclopedia of Philosophy*. New York: Free Press.

Fodor, J. 1968. *Psychological Explanation: An Introduction to the Philosophy of Psychology*. New York: Random House.

______. 1983. *The Modularity of Mind*. Cambridge: MIT Press.

Frankfurt, H. 1969. "Alternate Possibilities and Moral Responsibility." Reprinted in H. Frankfurt. 1988. *The Importance of What We Care About*. Cambridge: Cambridge University Press, pp. 1–10.

______. 1971. "Freedom of the Will and the Concept of a Person." Reprinted in H. Frankfurt. 1988. *The Importance of What We Care About*. Cambridge: Cambridge University Press, pp. 11–25.

______. 1982. "The Importance of What We Care About." Reprinted in H. Frankfurt. 1988. *The Importance of What We Care About*. Cambridge: Cambridge University Press, pp. 80–94.

______. 1988. *The Importance of What We Care About*. Cambridge: Cambridge University Press.

______. 1999. *Necessity, Violation and Love*. Cambridge: Cambridge University Press.

Freud, S. 1928. *The Future of an Illusion*. London: Hogarth Press and the Institute of Psychoanalysis.

______. 1977. *Introductory Lectures on Psychoanalysis*. Edited by James Strachey. New York: W. W. Norton.

Gardner, H. 1985. *The Mind's New Science: A History of Cognitive Revolution*. New York: Basic Books.

Gazzaniga, M., ed. 1995. *The Cognitive Neurosciences*. Cambridge: MIT Press.

Gibbard, A. 1990. *Wise Choices, Apt Feelings: A Theory of Normative Judgment*. Cambridge: Harvard University Press.

Gleick, J. 1987. *Chaos*. New York: Viking.

Goffman, E. 1959. *The Presentation of Self in Everyday Life*. New York: Doubleday and Company.

Goldman, A. 1969. "The Compatibility of Mechanism and Purpose." *Philosophical Review*, pp. 468–482.

______. 1970. *A Theory of Human Action*. Englewood Cliffs, N.J.: Prentice-Hall.

______. 1986. *Epistemology and Cognition*. Cambridge: Harvard University Press.

Goodman, N. 1978. *Ways of Worldmaking*. Indianapolis: Hackett Publishing.

Gould, S. J. 1980. *The Panda's Thumb*. New York: W.W. Norton.

______. 1999. *Rocks of Ages: Science and Religion in the Fullness of Life*. New York: Ballantine Publishing Group.

Gregory, R. L. 1987. *The Oxford Companion to the Mind*. Oxford: Oxford University Press.

Griffiths, P. E. 1997. *What Emotions Really Are: The Problem of Psychological Categories*. Chicago and London: University of Chicago Press.

Hacking, I. 1999. *The Social Construction of What?* Cambridge and London: Harvard University Press.

Hallie, P. 1979. *Lest Innocent Blood Be Shed: The Story of the Village of Le Chambon and How Goodness Happened There.* New York: Harper and Row.

Hameroff, S., A. W. Kasnziak, and A. Scott, eds. 1996. *Toward a Science of Consciousness: The First Tucson Discussions and Debates*. Cambridge and London: MIT Press.

Heidegger, M. 1962. *Being and Time (Sein und Zeit)*. English edition translated by J. Macquarrie and E. Robinson. New York: Harper & Row.

Heine, S. J., et al. 1999. "Is There a Universal Need for Positive Self-Regard?" *Psychological Review*, 106(4): 766–794.

Hill, T. 1991. *Autonomy and Self Respect*. New York: Cambridge University Press.

Hobbes, Thomas. 1651 [1999]. *Leviathan*. Oxford: Clarendon Press.

Hobson, J. A. 1999. *Consciousness*. New York: Scientific American Library.

Honderich, T., ed. 1995. *The Oxford Companion to Philosophy*. Oxford: Oxford University Press.

Howard, D. 1996. "The History That We Are: Philosophy as Discipline and the Multiculturalism Debate." In Anindita Balslev, ed. *Cross Cultural Conversation*. Atlanta: Scholars Press, pp. 43–76.

Hume, D. 1739 [1985]. *Treatise on Human Nature*. Middlesex, England: Penguin.

______. 1777 [1983]. *An Enquiry Concerning the Principles of Morals*. Edited by J. B. Schneewind. Indianapolis and Cambridge: Hackett Publishing.

Humphrey, N. 1992. *A History of the Mind*. New York: Simon & Schuster.

Humphrey, N., and D. C. Dennett. 1989. "Speaking for Ourselves." *Raritan: A Quarterly Review*, 9: 69–98.

Husserl, E. 1913. *Ideas*. English edition, 1972, translated by B. Gibson. New York: Collier Books.

Isen, A. M., and H. Levin. 1972. "Effect of Feeling Good on Helping: Cookies and Kindness." *Journal of Personality and Social Psychology*, 21: 384–388.

Ito, M., Y. Miyashita, and E. T. Rolls, eds. 1997. *Cognition, Computation, and Consciousness*. Oxford: Oxford University Press.

Jackendoff, R. 1987. *Consciousness and the Computational Mind*. Cambridge: MIT Press.

James, W. 1884. "The Dilemma of Determinism." Reprinted in John McDermott, ed. 1967. *The Writings of William James*. New York: Random House.

______. 1890. "The Stream of Thought." In W. James. *Principles of Psychology*. New York: Holt.

______. 1892. "The Stream of Consciousness." In W. James. 1961. *Psychology: The Briefer Course*. Edited by G. Allport. New York: Harper & Row.

Jones, E. E., and R. E. Nisbett. 1971. "The Actor and Observer; Divergent Perceptions of the Causes of Behavior." In E. Jones, D. Kanouse, H. Kelley, R. Nisbett, S. Valins, and B. Weiner, eds. *Attribution: Perceiving the Causes of Behavior*. Morristown, N.J.: General Learning Press.

Jopling, D. 2000. *Self-Knowledge and the Self*. New York: Routledge.

Juarrero, A. 1999. *Dynamics in Action: Intentional Behavior as a Complex System*. Cambridge and London: MIT Press.

Kagan, J. 1984. *The Nature of the Child*. New York: Basic Books.

Kagan, S. 1989. *The Limits of Morality*. Oxford: Clarendon Press.

Kane, R. 1989. "Two Kinds of Incompatibilism." Reprinted in T. O'Conner, ed. 1995. *Agents, Causes and Events: Essays on Indeterminism and Free Will*. New York: Oxford University Press, pp. 115–150.

______. 1996. *The Significance of Free Will*. New York: Oxford University Press.

Kant, I. 1964. *The Groundwork to the Metaphysics of Morals*. Translated by H. J. Paton. New York: Harper & Row.

Kekes, J. 1995. "Wisdom." In T. Honderich, ed. *The Oxford Companion to Philosophy*. Oxford and New York: Oxford University Press.

Kermode, F. 1967. *The Sense of an Ending: Studies in the Theory of Fiction*. New York: Oxford University Press.

Kim, J. 1993. *Supervenience and Mind: Selected Philosophical Essays*. Cambridge and New York: Cambridge University Press.

King, R. 1999. *Indian Philosophy: An Introduction to Hindu and Buddhist Thought*. Washington, D.C.: Georgetown Press.

Kitcher, P. 1986. *Vaulting Ambition*. Cambridge: MIT Press.

Koyré, A. 1957. *From a Closed World to an Infinite Universe*. Baltimore: Johns Hopkins University Press.

Kuhn, T. 1957. *The Copernican Revolution: Planetary Astronomy in the Development of Western Thought*. Cambridge: Harvard University Press.

______. 1962 (2nd ed., 1970). *The Structure of Scientific Revolutions*. Chicago: University of Chicago Press.

Kupperman, J. 1991. *Character*. New York and Oxford: Oxford University Press.

Lakoff, G., and M. Johnson. 1999. *Philosophy in the Flesh: The Embodied Mind and Its Challenge to Western Thought*. New York: Basic Books.

Laplace, P. S. 1951. *A Philosophical Essay on Probabilities*. Translated by F. W. Truscott and F. L. Emory. New York: Dover Publications.

Latané, B. 1981. "The Psychology of Social Impact." *American Psychologist*, 36: 343–356.

Latané, B., and J. Darley. 1970. *The Unresponsive Bystander: Why Doesn't He Help?* Englewood Cliffs, N.J.: Prentice Hall.

Latour, B., and S. Woolgar. 1979. *Laboratory Life: The Social Construction of Scientific Facts*. Beverly Hills, Calif.: Sage Publications.

Levine, J. 1983. "Materialism and Qualia: The Explanatory Gap." *Pacific Philosophical Quarterly*, 64: 345–361.

______. 2001. *Purple Haze: The Puzzle of Consciousness*. New York: Oxford University Press.

Leys, S. 1997. *The Analects of Confucius*. New York: W.W. Norton.

Little, D. 1991. *Varieties of Social Explanation: An Introduction to the Philosophy of Social Science*. Boulder: Westview Press.

Locke, J. 1690. *An Essay Concerning Human Understanding*. Annotated by A. C. Fraser, 1959. New York: Dover Publications.

Loewer, B. 1996. "Freedom from Physics: Quantum Mechanics and Free Will." *Philosophical Topics*, 24(2): 91–112.

London, P. 1970. "The Rescuers: Motivational Hypotheses About Christians Who Saved Jews from the Nazis." In J. R. Macaulay and L. Berkowitz, eds. *Altruism and Helping Behavior*. New York: Academic Press.

Lovejoy, A. O. 1936. *The Great Chain of Being: A Study of the History of an Idea*. Cambridge: Harvard University Press.

Luria, A. R. 1972. *The Man with a Shattered World* (reprinted 1987). Cambridge: Harvard University Press.

Lutz, C. 1988. *Unnatural Emotions*. Chicago: University of Chicago Press.

Lycan, W. G. 1987. *Consciousness*. Cambridge: MIT Press.

______, ed. 1990. *Mind and Cognition: A Reader*. Oxford: Blackwell.

______. 1996. *Consciousness and Experience*. Cambridge: MIT Press.

MacIntyre, A. 1959. "Hume on 'Is' and 'Ought.'" *Philosophical Review*, 68. Reprinted in W. D. Hudson, ed. 1969. *The Is–Ought Question*. London: Macmillan.

______. 1981. *After Virtue* (2nd ed., with postscript, 1984). Notre Dame, Ind.: Notre Dame University Press.

______. 1982. "How Moral Agents Became Ghosts: Or, Why the History of Ethics Diverged from That of the Philosophy of Mind." *Synthese*, 53: 295–312.

______. 1987. *Whose Justice? Which Rationality?* Notre Dame, Ind.: Notre Dame University Press.

______. 1999. *Dependent Rational Animals*. Peru, Ill.: Carus Publishing.

Malebranche, N. 1923. *Dialogues on Metaphysics and on Religion*. London: Allen & Unwin.

Marcel, A., and E. Basiach. 1988. *Consciousness in Contemporary Science*. Oxford and New York: Clarendon Press.

Marks, C. 1980. *Commissurotomy, Consciousness, and Unity of Mind*. Cambridge: MIT Press.

Marx, K. 1975. *The Economic and Philosophical Manuscripts of 1844*. New York: International Publishers.

May, L., M. Friedman, and A. Clark, eds. 1996. *Mind and Morals: Essays on Ethics and Cognitive Science*. Cambridge and London: MIT Press.

McGinn, C. 1989. "Can We Solve the Mind–Body Problem?" *Mind*, 97(891): 349–366.

______. 1991. *The Problem of Consciousness*. Cambridge: Blackwell

______. 1999. *The Mysterious Flame: Conscious Minds in a Material World*. New York: Basic Books.

McLaughin, B. 1992. "The Rise and Fall of British Emergentism." In A. Beckermann, H. Flohr, and J. Kim, eds. *Emergence or Reduction?: Essays on the Prospects of Nonreductive Physicalism*. Hawthorne, N.Y.: De Gruyter, pp. 49–93.

Mele, A. 1995. *Autonomous Agents: From Self-Control to Autonomy*. New York: Oxford University Press.

Merleau-Ponty, M. 1962. *Phenomenology of Perception*. Translated by C. Smith. New York: Humanities Press.

______. 1963. *The Structure of Behavior*. Translated by A. Fisher. Boston: Beacon Press.

Metzinger, T. 1995. *Conscious Experience*. Paderborn, Germany: Schöningh.

Metzinger, T., ed. 2000. *Neural Correlates of Consciousness: Empirical and Conceptual Questions*. Cambridge: MIT Press.

Milgram, S. 1974. *Obedience to Authority: An Experimental View*. New York: Harper &Row.

Nagel, T. 1979. "Moral Luck." In *Mortal Questions*. Cambridge: Cambridge University Press.

______. 1979. "What Is It Like To Be a Bat?" In *Mortal Questions*. Cambridge: Cambridge University Press.

______. 1986. *The View from Nowhere*. New York: Oxford University Press.

Nisbett, R. E., and L. Ross. 1980. *Human Inference: Strategies and Shortcomings of Social Judgement*. Englewood Cliffs, N.J.: Prentice Hall.

Nisbett, R. E., and T. D. Wilson. 1977. "Telling More Than We Can Know: Verbal Reports on Mental Process." *Psychological Review*, 84: 231–250.

Nisbett, R. E., C. Caputo, P. Legant, and J. Marecek. 1973. "Behavior as Seen by Actor and as Seen by Observer." *Journal of Personality and Social Psychology*, 27: 154–162.

Nozick, R. 1981. *Philosophical Explanations*. Cambridge: Harvard University Press.

______. 1981. "Philosophy and the Meaning of Life." In *Philosophical Explanations*. Cambridge: Harvard University Press, pp. 575–647.

Nussbaum, M. 1999. *Sex and Social Justice*. New York: Oxford University Press.

Nussbaum, M., and A. Sen, eds. 1992. *The Quality of Life*. New York: Oxford University Press.

Nussbaum, M. C. 1986. *The Fragility of Goodness: Luck and Ethics in Greek Tragedy and Philosophy*. New York: Cambridge University Press.

______. 1988. "Non-relative Virtues." In P. A. French, T. E. Uehling, and H. K. Wettstein, eds. *Midwest Studies in Philosophy. Vol. 13. Ethical Theory: Character and Virtue*. Notre Dame, Ind.: Notre Dame University Press.

______. 1994. *The Therapy of Desire: Theory and Practice in Hellenistic Ethics*. Princeton: Princeton University Press.

______. 2000. *Women and Human Development: The Capabilities Approach*. Cambridge: Cambridge University Press.

O'Conner, T., ed. 1995. *Agents, Causes and Events: Essays on Indeterminism and Free Will*. New York: Oxford University Press.

O'Conner, T. 2000. *Persons and Causes*. New York: Oxford University Press.

Oliner, S. P., and P. M. Oliner. 1988. *The Altruistic Personality: Rescuers of Jews in Nazi Europe*. New York: Free Press.

Parfit, D. 1971. "Personal Identity." *Philosophical Review*, 80(3): 27.

______. 1984. *Reasons and Persons*. Oxford: Oxford University Press.

Peitot, J., F. Varela, B. Pachoud, and J. Roy. 1999. *Naturalizing Phenomenology: Issues in Contemporary Science*. Stanford: Stanford University Press.

Penrose, R. 1994. *Shadows of the Mind: A Search for the Missing Science of Consciousness*. New York: Oxford University Press.

Pepper, S. 1942. *World Hypotheses*. Berkeley: University of California Press.

Perry, J., ed. 1975. *Personal Identity*. Berkeley: University of California Press.

______. 2001. *Knowledge, Possibility and Consciousness*. Boston: MIT Press.

Pickering, A. 1984. *Constructing Quarks: A Sociological History of Particle Physics*. Chicago: University of Chicago Press.

Pinker, S. 1995. *The Language Instinct*. New York: Harper Collins.

______. 1997. *How the Mind Works*. New York: W.W. Norton.

Place, U. T. 1956. "Is Consciousness a Brain Process?" Reprinted in V. C. Chappel, ed. 1981. *The Philosophy of the Mind*. First printed by Prentice Hall in 1962. New York: Dover Publications, pp. 101–109.

Polkinghorne, J. 1998. *Belief in God: In an Age of Science*. New Haven and London: Yale University Press.

Purves, D., et al. 1997. *Neuroscience*. Sunderland, Mass.: Sinauer Associates.

Putnam, H. 1963. "Brains and Behavior." Reprinted in N. Block, ed. 1980. *Readings in Philosophy of Psychology*, vol. 1. Cambridge: Harvard University Press, pp. 24–36.

Quine, W. V., and J. Ullian. 1970. *The Web of Belief*. New York: Random House.

Rawls, J. 1971. *Theory of Justice*. Cambridge: Harvard University Press.

Revel, J. F., and M. Ricard. 1999. *The Monk and the Philosopher: A Father and Son Discuss the Meaning of Life*. Translated by J. Canti. New York: Schocken Books.

Rorty, R. 1989. *Contingency, Irony, and Solidarity*. Cambridge: Cambridge University Press.

Rosenberg, A. 1988. *Philosophy of Social Science*. Boulder: Westview Press.

______. 2000. *Philosophy of Science: A Contemporary Introduction*. London: Blackwell.

______. 2000. *Darwinism in Philosophy, Social Science and Policy*. Cambridge: Cambridge University Press.

Rosenthal, D. M., ed. 1991. *The Nature of Mind*. New York: Oxford University Press.

Ross, L. 1977. "The Intuitive Psychologist and His Shortcomings: Distortions in the Attribution Process." In L. Berkowitz, ed. *Advances in Experimental Social Psychology*. New York: Academic Press.

______. 1988. "Situationist Perspectives on the Obedience Experiments: Review of A. G. Miller (1986)." *Contemporary Psychology*, 33: 101–104.

Ross, L., T. M. Amabile, and J .L. Steinmetz. 1977. "Social Roles, Social Control, and Biases in Social Perception Processes." *Journal of Personality and Social Psychology*, 35: 485–494.

Ross, L., M. R. Lepper, and M. Hubbard. 1975. "Perseverance in Self Perception and Social Perception: Biased Attributional Processes in the Debriefing Paradigm." *Journal of Personality and Social Psychology*, 32: 880–892.

Ruse, M. 1996. *The Concept of Progress in Evolutionary Biology*. Cambridge: Harvard University Press.

Russell, R. J., N. Murphy, T. C. Meyering, and M. A. Arbib, eds. 1999. *Neuroscience and the Person: Scientific Perspectives on Divine Action*. Vatican City State and Berkeley: Vatican Observatory Publications and Center for Theology and the Natural Sciences.

Ryle, G. 1949. *The Concept of Mind*. Chicago: University of Chicago Press.

Sacks, O. 1985. *The Man Who Mistook His Wife for a Hat and Other Clinical Tales*. New York: Summit.

Scheffler, S. 1982. *The Rejection of Consequentialism*. New York: Oxford University Press.

______. 1992. *Human Morality*. New York and Oxford: Oxford University Press.

Schoeman, F., ed. 1987. *Responsibility, Character, and the Emotions: New Essays in Moral Psychology*. Cambridge: Cambridge University Press.

Searle, J. 1980. "Minds, Brains, and Programs." *Behavioral and Brain Sciences*, 3: 417–424.

______. 1984. *Minds, Brains and Science*. Cambridge: Harvard University Press.

______. 1992. *The Rediscovery of the Mind*. Cambridge and London: MIT Press.

Sellars, W. 1963. *Science, Perception, and Reality*. London: Humanities Press.

Sen, A. 1980. "Equality of What?" In S. McCurrin, ed. *Tanner Lectures on Human Values*, vol. 1. Cambridge: Cambridge University Press.

Shallice, T. 1988. *From Neuropsychology to Mental Structure*. Cambridge: Cambridge University Press.

Shoemaker, S., and R. Swinburne. 1984. *Personal Identity*. Oxford: Blackwell.

Shweder, R., and R. LeVine, eds. 1984. *Culture Theory: Essays on Mind, Self, and Emotion*. New York: Cambridge University Press.

Siewert, C. 1998. *The Significance of Consciousness*. Princeton: Princeton University Press.

Singer, Peter. 1979. *Practical Ethics*. New York: Cambridge University Press.

Skilton, A. 1994. *A Concise History of Buddhism*. Birmingham, Ala.: Windhouse Publications.

Skinner, B. F. 1953. *Science and Human Behavior*. New York: Macmillan.

______. 1971. *Beyond Freedom and Dignity*. New York: Alfred A. Knopf.

______. 1974. *About Behaviorism*. New York: Alfred A. Knopf.

______. 1975. *Verbal Behavior*. New York: Appleton-Century-Crofts.

Skyrms, B. 1996. *Evolution of the Social Contract*. Cambridge: Cambridge University Press.

Smart, J. J. C. 1959. "Sensations and Brain Processes." *Philosophical Review*, 68: 141–156.

Snow, C. P. 1969. *Two Cultures and the Scientific Revolution*. London: Cambridge University Press.

Sober, E., and D. S. Wilson. 1998. *Unto Others: The Evolution and Psychology of Unselfish Behavior*. Cambridge: Harvard University Press.

Spence, D. 1982. *Narrative Truth and Historical Truth: Meaning and Interpretation in Psychoanalysis*. New York: W.W. Norton.

Stephens, G. L., and G. Graham. 2000. *When Self-Consciousness Breaks: Alien Voices and Inserted Thoughts*. Cambridge: MIT Press.

Stern, D. 1985. *The Interpersonal World of the Infant: A View from Psychoanalysis and Developmental Psychology*. New York: Basic Books.

Strawson, G. 1994. *Mental Reality*. Cambridge: MIT Press.

Strawson, P. F. 1962. "Freedom and Resentment." Reprinted in G. Watson, ed. 1982. *Free Will*. Oxford and New York: Oxford University Press, pp. 59–80.

Taylor, C. 1977. "Responsibility for Self." Reprinted in G. Watson, ed. 1982. *Free Will*. Oxford and New York: Oxford University Press, pp. 111–126.

______. 1989. *Sources of the Self: The Making of the Modern Identity*. Cambridge: Harvard University Press.

Taylor, S. E. 1989. *Positive Illusions: Creative Self-Deception and the Healthy Mind*. New York: Basic Books.

Taylor, S. E., and J. Brown. 1988. "Illusion and Well-Being: A Social Psychological Perspective on Mental Health." *Psychological Bulletin*, 103: 193–210.

Tec, N. 1986. *When Light Perceived the Darkness: Christian Rescue of Jews in Nazi-Occupied Poland*. Oxford: Oxford University Press.

Tye, M. 1995. *Ten Problems of Consciousness: A Representational Theory of the Phenomenal Mind*. Cambridge: MIT Press.

van Inwagen, P. 1983. *An Essay on Free Will*. Oxford: Clarendon Press.

Varela, F., E. Thompson, and E. Rosch. 1991. *The Embodied Mind: Cognitive Science and Human Experience*. Cambridge and London: MIT Press.

Wallace, A. 1993. *Tibetan Buddhism from the Ground Up*. Boston: Wisdom Publications.

______. 2000. *The Taboo of Subjectivity: Toward a New Science of Consciousness*. New York: Oxford University Press.

Watson, G. 1975. "Free Agency." Reprinted in G. Watson, ed. 1982. *Free Will*. Oxford and New York: Oxford University Press, pp. 96–110.

Watson, G., ed. 1982. *Free Will*. Oxford and New York: Oxford University Press.

Weinberg, S. A. 1999. "A Designer Universe?" *The New York Review of Books*, 46(16): 46–47.

Wilkes, K. 1988. *Real People: Personal Identity Without Thought Experiments*. New York: Oxford University Press.

Williams, B. 1981. "Moral Luck." In B. Williams. *Moral Luck: Philosophical Papers*. Cambridge: Cambridge University Press.

______. 1981. "Persons, Character, and Morality." In B. Williams. *Moral Luck: Philosophical Papers*. Cambridge: Cambridge University Press.

______. 1985. *Ethics and the Limits of Philosophy*. Cambridge: Harvard University Press.

Wilson, E. O. 1975. *Sociobiology: The New Synthesis*. Cambridge and London: Harvard University Press.

______. 1978. *On Human Nature*. Cambridge and London: Harvard University Press.

Winch, P. 1958. *The Idea of a Social Science*. New York: Humanities Press.

Wolf, S. 1982. "Moral Saints." *Journal of Philosophy*, 79: 419–439.

______. 1990. *Freedom Within Reason*. New York: Oxford University Press.

Wollheim, R. 1999. *On the Emotions*. New Haven and London: Yale University Press.

Wong, D. B. 1984. *Moral Relativity*. Berkeley: University of California Press.

Wright, R. 1994. *The Moral Animal: The New Science Of Evolutionary Psychology*. New York: Random House.

Zinsser, H. 1934. *Rats, Lice and History: A Chronicle of Pestilence and Plagues*. New York: Black Dog and Leventhal Publishers.

Zukav, G. 1989. *The Seat of the Soul*. New York: Simon & Schuster.

ACKNOWLEDGMENTS

This book is the result of two strokes of good *karma*. First, I was greatly honored to be named "Romanell Phi Beta Kappa Professor" in 1998–1999 by the National Office of Phi Beta Kappa in Washington, D.C., for distinguished contributions to philosophy and to the public understanding of philosophy. This book is a significant expansion of the three Romanell Lectures on "Science and the Human Image" that I gave at Duke University in October of 1999 in my capacity as Romanell Professor. Second, in the summer of 1999, I was invited by the Mind and Life Institute in Boulder, Colorado, to participate in a small conference in March of 2000 in Dharamsala, India, with several other Western intellectuals, mostly scientists, and His Holiness, the 14th Dalai Lama. The topic was "Destructive Emotions." His Holiness is committed to promoting a universal secular ethic, and he seeks to incorporate Western philosophical and scientific ideas about the nature of mind and persons into Buddhism. *Dharma*, the Buddhist Way, is self-consciously humble about its grasp of reality and is thus open to adjusting and completing itself, as the truth requires, in a way that, in my experience, is resisted by the entrenched perennial philosophy of the West.

Besides providing me with occasions to write and present new material, both invitations required me to write and speak in a different way than I am accustomed. Normally, I write for and speak to fellow philosophers. But Phi Beta Kappa asked me to prepare lectures for a general intellectual audience. And the Mind and Life Institute asked me to introduce His Holiness to Western ways of thinking about the role of the emotions in moral life in a way that did not assume any knowledge of Plato, Aristotle, Hobbes,

Rousseau, Hume, or Kant. So my job in both cases was to speak about matters of philosophical importance without oversimplifying but, at the same time, without being able to avail myself of the specialized idiom I use when speaking to, and writing for, members of my own academic tribe. This book is the result.

I have been thinking about this problem since I was a teenager. Thus my deepest gratitude goes to all my friends and family members, students and colleagues who have puzzled with me over the years on the problem of the conflict between the humanistic and scientific images.

William Frucht, Senior Editor at Basic Books, was enthusiastic about the book from the start, and taught me many things about good writing. Thanks also to Maggie Mauney, Vanessa Mobley, Kay Mariea, and John Thomas for their very professional assistance in preparing the final manuscript.

The book was written in a variety of places: Durham, North Carolina; Boston and Cambridge, Massachusetts; Delhi and Dharamsala, India; Jacó and San José, Costa Rica; Paris, France; Athens, Greece; Rethymno, Crete; Oslo and Vestby, Norway; San Francisco; and Scarborough, Maine. Different places and the people in them inspire me in different ways. I am grateful to the people and the places.

INDEX